Eco-Concepts

ECOCRITICAL THEORY AND PRACTICE

Series Editor: Douglas A. Vakoch, METI

Ecocritical Theory and Practice highlights innovative scholarship at the interface of literary/cultural studies and the environment, seeking to foster an ongoing dialogue between academics and environmental activists.

Recent Titles

Eco-Concepts: Critical Reflections in Emerging Ecocritical Theory and Ecological Thought, edited by Cenk Tan and İsmail Serdar Altaç

Contemporary Ecocritical Methods, edited by Camilla Brudin Borg, Jørgen Bruhn, and Rikard Wingård

Post Green: Literature, Culture, and the Environment, edited by Murali Sivaramakrishnan and Animesh Roy

Multispecies Thinking in the Classroom and Beyond: Teaching for a Sustainable Future, edited by Patty Born

Intermedial Ecocriticism: The Climate Crisis Through Art and Media, by Jørgen Bruhn and Niklas Salmose

Ibero-American Ecocriticism: Cultural and Social Explorations, edited by J. Manuel Gómez

Animal Texts: Critical Animal Concepts for American Environmental Literature, by Lauren E. Perry-Rummel

Eco-Concepts

Critical Reflections in Emerging Ecocritical Theory and Ecological Thought

Edited by Cenk Tan and İsmail Serdar Altaç

LEXINGTON BOOKS
Lanham • Boulder • New York • London

Published by Lexington Books
An imprint of The Rowman & Littlefield Publishing Group, Inc.
4501 Forbes Boulevard, Suite 200, Lanham, Maryland 20706
www.rowman.com

86-90 Paul Street, London EC2A 4NE

British Library Cataloguing in Publication Information Available

Library of Congress Cataloging-in-Publication Data Available

ISBN 978-1-66692-348-3 (cloth)
ISBN 978-1-66692-349-0 (electronic)

Contents

Editors' Introduction

Cenk Tan & İsmail Serdar Altaç

Despite the long-standing history of unsustainable practices of human beings, which might be dated back to the Industrial Revolution, not until the 1970s did these practices and their planetary impact become an officially recognized international issue. In response to the issue's rising importance, the United Nations convened the "Environmental Conference" in Stockholm in 1972. The United Nations Environment Programme (UNEP) was established as an essential result of this conference whose first day, June 5, was proclaimed "World Environment Day." However, over the past half-century, environmental problems have steadily increased. Crowded cities, growing population, polluted air and water, extreme weather events, degradation of soil, deforestation, habitat loss, endangered species, deteriorating health, floods, droughts, pandemics, etc., are on the rise. In a 2010 work, Bill McKibben, author, educator, environmentalist, and the founder of the 350.org movement, emphasized that the Earth has changed and thus advocated naming the new planet "Eaarth." In McKibben's words: "We've turned our sweet planet into Eaarth, which is not as nice. We're moving quickly from a world where we push nature around to a world where nature pushes back—and with far more power. But we've still got to live on that world, so we better start figuring out how" (101).

In this half-century period marked by adversities, can we talk about any achievements? The most significant gain is the realization and acceptance that those who have haphazardly exploited the Earth's resources are the ones responsible for damaging their own environment. The naming of this new era as the "Anthropocene" is the most distinct reflection of this awareness and acceptance. In essence, awareness of human power to alter the environment was expressed in the Stockholm Declaration (1972):

> In the long and tortuous evolution of the human race on this planet a stage has been reached when, through the rapid acceleration of science and technology, man has acquired the power to transform his environment in countless ways and

on an unprecedented scale. (Declaration of the United Nations Conference on the Human Environment)

Being just another constituent of the planet and having a biological agency, human beings naturally exert an influence on the environment, but what's ethically problematic is the unprecedented temporal and spatial scale of this influence (Callicott 215). With this awareness and acceptance of humanity's destructive power, green technologies and science-based solutions for a sustainable planet are growing every day. However, environmentally conscious writers, bioregional thinkers, philosophers, and environmental critics who trigger emotional and ideological transformations by establishing an eco-language play a significant role in the ongoing change. Through this eco-language containing numerous eco-concepts, our language has been enriched, and the language we speak has become more eco-conscious.

In this book, we have compiled these new eco-concepts which not only strengthen our connections with the world but also remind us of our responsibility to the Earth. These eco-concepts, products of a collective effort in the face of the commercialization of nature, herald a healthier future. They also accelerate the long-awaited Ecological Age with a sense of yearning. Consisting of 21 chapters from international scholarship across the globe, *Eco-Concepts: Critical Reflections in Emerging Ecocritical Theory and Ecological Thought* is committed to identifying and discussing the latest and most prevalent notions in ecocritical and ecological thought. This volume provides a critical overview of environmental concepts that have been on the agenda during the last two decades.

Part I of the volume centers on "Ecological Concepts" which stand out as major components that act as driving forces in the human and non-human spheres. In the first chapter, Cenk Tan defines, rethinks and calls into question "Ecocide" as one of the most problematic and grave ecological outcomes of the Anthropocene. Tan highlights that the struggle against ecocide is a challenging one that, despite all odds, must be continued and accomplished with the ultimate recognition of ecocide as an international law. Following ecocide, İsmail Serdar Altaç questions the discourses shaped by "Petroculture" and examines them within the larger context of utopian thought to find a new vein of interpretation among them. The next part is written by Clara Soudan who advocates that "Dark Ecology" reaches into an unconquered universe that eludes current ontology and epistemology, making a significant addition to ecocritical philosophy. In this context, Rebekah A. Taylor-Wiseman touches upon the significance of the "Anthropocene and Capitalocene" which have profoundly influenced the environmental humanities. Wiseman contends that despite its flaws, the Anthropocene has gained widespread acceptance and though it may appear insufficient as an eco-concept, it is usefully imperfect, since even critiques like Moore's allow debates that uncover injustices and advocate greater responsible management of the Earth's resources. Complementing the Anthropocene and Capitalocene are the eco-concepts "Plantationocene and Chthulucene" which are revisited in detail by Lenka Filipova. The Plantationocene and Chthulucene are critical for understanding the environmental consequences of colonialism. These notions reflect historical and present injustices and provide a viewpoint that perceives the world as a dynamic interplay of human and non-human actors. The subsequent chapter in this part centers on "Bioregionalism and Biocultural Region." Abhra Paul and Amarjeet Nayak suggest that Bioregionalism aims to address ecological challenges at the local level. However, the current priority is to develop a more globally sustainable world in order to address anthropogenic challenges. As a result, environmental issues may be addressed locally while keeping ecological, social, cultural, and political limits in mind and such actions will be beneficial to international sustainability. Next, in chapter 7, Brian Deyo's "Reading Ecophobia in the Capitalocene" examines the eco-concept "Ecophobia" and sets forth the relationship between the environment, culture, climate and capitalism. In the succeeding chapter, Philippe Lynes, who looks into "Eco-Deconstruction," defends that the eco-concept, like environmental posthumanities, confronts an unaddressed gap between the politics of human environmental justice and a broader, ethical obligation to nonhumans and their ecosystems. The eco-deconstruction chapter is followed by Erden El's analysis of "Environmental Justice." El demonstrates "the inequities class and ethnicity" as the departing point of Environmental Justice and lays bare the required reformative measures. Subsequently, Panchali Bhattacharya and Pritam Panda draw attention to the potential for infusing indigenous ecosophy into the study of a rising eco-concept known as "Ecopsychology," which can significantly improve the mental stability essential for well-being. In the last chapter of this part, Ana Simić outlines and stipulates the recent eco-concept of "Hyperobjects" which she argues to have played a crucial function in environmental ethics by stressing the interdependence of ecological systems and recognizing the constraints of human perception and representation.

Part II of *Eco-Concepts* navigates through "Ecocriticisms" consisting of the most recent fields that have flourished from ecocritical scholarship. In the chapter titled, "Restoration Ecocriticism," Ufuk Özdağ postulates that literature scholars are more required than ever before to communicate skills and expertise in land healing, to disseminate restoration research and practices from throughout the world, and to spark grassroots restoration efforts. Restoration ecocriticism might become one of their tools for raising awareness, healing wounds, and mobilizing community participation in landscape-scale Earth restoration. Subsequent to this chapter is Denis Petrina's chapter on "Affective Ecocriticism" which investigates a range of negative affective reactions to the environmental problems, ranging from passive and indifferent material and mental exhaustion to more "militant" environmentalist killjoy. This chapter is followed by Milena Škobo's overview of "Empirical Ecocriticism" which has elevated our knowledge of literature's potential to shape environmental awareness to a whole new level, providing a unique and evidence-based viewpoint on the transformational influence of environmental narratives. Škobo demonstrates the ability of empirical research methods to open up new avenues in ecocriticism as well as its possible limitations. Next, Gina Stamm, in her chapter on "Material Ecocriticism" affirms that materialist ecocritical reading has the advantage of drawing scholarly attention to nonhuman nature's narrative and agential force. By taking these depictions seriously, it gives authors of literature credit for their own "readings" of nature. Finally, the ecocriticism section of the book rounds up with Mahinur Gözde Kasurka, who draws an inquisitive portrait of "Posthuman Ecocriticism" and affirms that it functions as an eco-concept that allows us to conceive affirmative horizons in environmental humanities by stressing the existence of two opposing conditions at the same time: to begin, it is impossible to overlook the shattered presence of the surroundings. Second, we may still hold to life by learning more from nature.

Part III sets off to traverse disciplinary boundaries and takes into consideration the essential denominations in the environmental sphere. In her chapter on "Blue Humanities," Aina Vidal-Pérez depicts the ways in which the study of nautical literature offers new scales, lexica, tropes, narratives, and figures, as well as other poetic, formal, and overall qualities. The author points out that a study of ocean poetics and how they depict the complexities of the environmental issue might aid in determining how these views are translated to the social arena. This is followed by Inna Sukhenko who envisions perspectives into the "Nuclear Humanities" which the scholar debates and delineates as a complex and multi-faceted interdisciplinary field. Next is the "Critical Animal Studies" chapter where Ziba Rashidian explores the rise and growth of the subfield by elaborating that CAS's struggle with anthropocentrism exemplifies a major difficulty to our articulation of eco-concepts appropriate for dealing with the Anthropocene and whatever dispensation follows. In the next chapter of *Eco-Concepts*, "Unravelling Critical Plant Studies," Kübra Vural Özbey indicates that the concept, driven by ecocritical theory, seeks to resolve complications between humans and plants via the establishment of a discourse founded on appreciation, care, and respect. The concluding chapter of *Eco-Concepts* is offered by Narayan Jena, and it scrutinizes "Gandhian Ecosophy." The author maintains that Gandhi's ecophilosophy transcends time and space. It has worldwide ramifications as an eco-concept and pervades both episteme and actions addressing global ecology. His doctrines of *ahimsa* (nonviolence) and *aparigraha* (non-possession) serve as ethical principles for nations that believe in the ideology of progress through immense technological development.

In closing, we anticipate that *Eco-Concepts: Critical Reflections in Emerging Ecocritical Theory and Ecological Thought* will offer a positive contribution to environmental humanities and ecological awareness. As the contributors, we sincerely share the aspiration that the Ecological Age will push forward in order to make progress and finally be able to restore and compensate all that has been deemed lost and/or severely damaged with the anthropocene.

REFERENCES

Callicott, J. Baird. "The Land Ethic." *A Companion to Environmental Philosophy*. Edited by Dale Jamieson. Blackwell, 2001, pp. 204–217.

"Declaration of the United Nations Conference on the Human Environment." *University of Minnesota Human Rights Library*, 1972, hrlibrary.umn.edu/instree/humanenvironment.html. Accessed December 10, 2023.

McKibben, Bill. *Eaarth: Making a Life on a Tough New Planet*. Times Books Henry Holt and Company, 2010.

PART I

Reimagining Ecological Concepts

Chapter 1

Revisiting Ecocide

At the Threshold of International Ecocide Law

Cenk Tan

Coined in the 1970s, ecocide is roughly defined as immense devastation inflicted on the natural environment. Despite its late coining, ecocide has existed throughout the ages. However, the emergence of the concept has contributed to the acknowledgement of this fact, causing ecological awareness around the globe. Nonetheless, this awareness has not been able to stop ecocidal activities. Inceptive efforts of activists such as Polly Higgins have managed to come a long way in this struggle which at all cost must carry on and be progressed until a concrete result is achieved.

The traces of ecocide may be tracked in literary works from around the world. Most of these works are classified under utopian/dystopian fiction, climate fiction and science fiction. Joseph Conrad's *Heart of Darkness*, J.G. Ballard's *The Drought* and *The Drowned World*, Octavia Butler's *The Parable of the Sower*, Margaret Atwood's *Oryx and Crake*, Brian Aldiss's *Hothouse*, Frank Herbert's *Dune*, Kim Stanley Robinson's *New York 2140* and *Children of Men* by P.D. James are some of the novels in English and American literatures where ecocide is dealt with.

This chapter explores ecocide as an eco-concept from its origin to its current state with selected cases from around the globe. It goes without saying that ecocide is one of the most tangible, urgent and substantial eco-concepts that has been on the agenda over the last two decades. Despite numerous efforts from scholars and environmentalists, ecocide has not yet been recognized as an international crime. The environmentalist community has not managed to push for an international law that condemns and punishes all those that engage in ecocidal activities. Therefore, the chapter asserts that ecocide and ecocidal tendencies must be put on the agenda in order to make it more visible and maintain the issue's actuality. Moreover, it also posits that campaigns have to be revitalized and supported to put pressure on international organizations such as the UN, so as to push forward for the ultimate aim of obtaining an international ecocide law that will act as a deterrent against all individual and corporate entities who pursue their monetary interests. The chapter highlights that the international ecocide law is a moral obligation and unless it is established, ecocidal activities will proceed terrorizing natural areas around the globe. Therefore, it proposes to take further initiative to push the authorities towards the urgent legislation of international ecocide law.

Ecocide is a term obtained from the Greek, *oikos* meaning house, home and Latin *caedere* meaning to destroy and kill (Mahta and Mertz 4). The term was coined by Arthur W. Galston in 1970 who suggested using this term to describe the disastrous effects of using the herbicide Agent Orange during the Vietnam War (Gardashuk 2). In this war, the Americans used 57,000 tons of herbicides (Agent Orange) and 23,000 tons of defoliants (13 formulations), destroying 17 million acres of forest and dropping approximately 7.9 million bombs (Govorushko 292). After the Vietnam War, the Gulf War proved to be the second major example of (military) ecocide with 550 oil wells burned and oil spills transmitted over 2500 km^2 of land (296).

In his speech to the United Nations Stockholm Conference on the Human Environment in 1972, Swedish Prime Minister Olof Palme introduced the term *ecocide* to allude to the effects of the Vietnam War (2). In its most commonly used context, ecocide refers to the destruction of natural areas. Various scientists and government officials have issued calls for ecocide to be accepted a war crime (5). In 2013, *The End Ecocide Movement* led by European citizens demanded the EU to maintain a regulation to interdict ecocide which attracted attention as it was advocated by 175,000 European citizens (5). Scottish barrister and environmental activist Polly Higgins defined the ecocide as "extensive damage to, destruction of or loss of ecosystem(s) of a given territory whether by human agency or by other causes, to such an extent that peaceful enjoyment by the inhabitants of that territory has been or will be severely diminished" (Higgins 16). Thus, ecocide directly affects the natural environment as well as the citizens living in or around that particular area. However, whether or not responsibility arises when damage to natural areas was not meant but came to being as a consequence of industrial activity is another matter of dispute (Mahta and Mertz 5).

Furthermore, various theories have been put forward concerning the origin of ecocide. According to environmental sociologist Franz J. Broswimmer, the very first phase in this origin goes back 60,000 years ago when Homo sapiens developed language which led to the birth of culture and innovation (9).

The second phase involves the creation of agriculture some 10,000 years ago which caused a serious destruction of megafauna resulting in "mass extinction combined with climatic and demographic changes to produce the food crisis in prehistory" (Broswimmer 9). The third and most crucial phase in the origin of ecocide was the emergence of modernity characterized by three major determinants, "the increasing division of labour, the capitalist mode of production, and the emergence of the modern nation-state" (10). Thus, the free-market economy was established and acknowledged as a global system around the world, ultimately leading to exploitation of the environment and its rationalization (10). Finally, "the exploitation of nature was universalized and commodified. In the end, the imperatives of late modernity produced the global framework in which ecocidal tendencies greatly accelerated" (10).

Within the scope of these three phases, it is the third phase that had the most profound impact on Earth in terms of anthropogenic climate change. In modernity, the grave issue of ecocide has evolved from a local problem to a global matter, and this was characterized by four distinct qualities: "the increasing division of labour, the capitalist mode of production, the emergence of the modern nation-state and the process of colonization" (Vorster 871). Thus, personal entrepreneurship and mercantile rivalry were valued as the useful driving force of advancement and illumination (871). The capitalist system envisaged a model society which promised people the freedom to buy, sell and manufacture the maximum number of goods and as a consequence the drive for maximum production in turn led to the plundering of other unspoiled areas of Earth (871). Hence, colonization became a crucial motive in the emergence of ecocide. A particular example of the effect of colonization on ecocide is the significant decrease of buffalo in North America which eventually resulted in the boost of trade of fur and beaver to Europe (Parent 324). The dramatic extinction of buffalo was supported by Roman Catholic colonizers as they thought it would disable the mobilization of their subjects (330). After the fur trade, a major commodity for Britain became lumber which was mainly imported from Canada to be used in ships and in 1808, ten ships loaded with lumber were transported to Britain every day (Wilson 126). Other examples concerning colonialism related destruction of nature are "commercial whaling, ecological destruction coupled with mining activities in colonies, the worldwide sugar industry, deforestation and pollutions of rivers and lakes" (Vorster 871). In addition, modern warfare also had a critical impact on ecocide as the two World Wars have surpassed all other wars in terms of the scale of destruction and number of deaths (872). Current globalist tendencies continue to promote modern ecocide as globalism inevitably causes more impoverishment and leads to the deterioration of the ecosystem (872).

As a consequence, three basic forms of ecocide have emerged throughout history: military, colonial and industrial ecocide. Peter Hough states that military ecocide represents the damage of the natural environment while fighting or preparing for war which has a long history and is still a common aspect of modern conflicts (Hough 10). Despite all efforts by environmental activists and scientists around the globe, military conflicts continue to inflict a heavy toll on nature. The second form is colonial ecocide which explores colonial invasions and ecocide throughout the long sixteenth century that paved the way for the creation of London's mega-mining firms during Europe's industrialization in the nineteenth century (Selwyn 127). Continuous mass extinction events, such as indigenous genocides and chattel slavery have altered global landscapes, atmospheres, and biophysical processes, forming the foundation of racial capitalism (127). Finally, the third form is industrial ecocide which exposes industry's endless ecocidal tendencies not to mention the deforestation for agriculture and urbanization that results from industrial product manufacturing. Hence, the course of industrial civilization continues to be ecocidal (Dunlap and Brock 3). Whatever the human intent and activity is, the ecocidal results for the natural environment are catastrophic. Alan Sangster specifies urban smog, acid rain, chlorofluorocarbons and heat waves as concrete evidence of ecocide (103–110).

Moreover, ecocide occurs in two basic types, produced by human activity or come to exist as a result of calamity (Higgins 16–17). Human caused ecocide is directly affiliated with corporatism and corporatist action whereas natural ecocide includes natural catastrophes such as earthquakes, floods and tsunamis which cause serious breakdowns of the ecosystem (17). Higgins describes the term ecocide as: "A value system based on a lack of regard for all life now needs to be universally outlawed as well. Kill our planet and we kill ourselves. Ecocide is death by a thousand cuts: each day the life-source which feeds and nourishes our human life is damaged and destroyed a little more" (17). The environmental activist highlights that the destruction of the planet is a universal crime as Earth harbours life for all living beings including fauna and flora. To that end, humans are not the only ones that ought to be concerned about the ecological genocide. Nevertheless, as humans are a part of the ecosystem, they are inevitably bound to be affected by all types of ecocide. Therefore, ecocide will have devastating impacts on future human generations since they will be obliged to live with the consequences of this environmental catastrophe. For these reasons, Higgins (1968–2019) was an active advocate for the establishment of the Law of ecocide which maintains the recognition of ecocide as an international crime (Higgins *Dare to Be Great*, 124). Following Higgins's tragic early death in April 2019, the movement to establish the Law of ecocide was renamed the *Stop ecocide campaign* (*Dare to Be Great*, 118). Section 6 of the ecocide Act, proposed to the UN Law Commission defends that: "The right to life is a universal right and where a person, company organization, partnership, or any other legal entity causes extensive damage to, destruction of or loss of human and or non-human life of the inhabitants of a territory . . . is guilty of the crime of ecocide" (124).

Thus, the establishment of an ecocide Law as international law will forbid large scale destruction of natural areas and ensure the protection and preservation of all people and living beings negatively affected by the ecocide (125). Nonetheless, before the establishment of the Ecocide Law, it is imperative to distinguish different forms of ecocide. In an article published by Polly Higgins, Damien Short and Nigel South, it is purported that environmental crimes can be classified into primary and secondary types ("Protecting the Planet," 252). Those that come to exist as a direct consequence of environmental destruction or exploitation are considered primary environmental crimes whereas those that manifest as symbiotic green harms resulting from the conditions that emerge as a result of the exploitation of environmental damages and/or crises are viewed as secondary crimes (252). In specific, primary crimes include "crimes/harms of water pollution, crimes/harms of air pollution, crimes/harms of deforestation and spoiling of the land and crimes/harms against animals/non-human species" (252–253). Nonetheless, Higgins argues that responses to these primary environmental crimes usually fall short and government authority's administration turns out to be insufficient to enforce measures and implement penalization against the responsible fraction (254). To that end, strict measures and restrictions need to be applied in order to provide a response for grave environmental issues such as climate change, excessive carbon emission and irreversible damage to the natural sphere (256). Cullinan contemplates that "a primary cause of environmental destruction is the fact that current legal systems are designed to perpetuate human domination of nature instead of fostering mutually beneficial relationships between humans and other members of the earth community" (144). Thus, it is obvious that more initiatives will need to be taken from the governmental perspective to successfully halt and reverse the condition of ecocide. This will need to be accomplished on a national scale initially, but imperatively on the international scale at large. So far, only ten nations have included ecocide in their penal codes with Vietnam as the forerunner in 1990 and the others are former Soviet countries, namely: Russian Federation, Kazakhstan, Kyrgyzstan, Georgia, Belarus, Tajikistan, Armenia, Moldova and Ukraine (Sarliève 234). However, these national penal codes are far from complying with international standards and efficacy.

The foremost problem with ecocide is the very fact that there exists no international, uniform law that underlines and protects the right of existence of all natural areas and natural life forms. Despite ongoing efforts by the

UN Conference on the Human Environment (1972) where Richard A. Falk proposed a draft of an 'International Convention on the Crime of Ecocide' in 1973 as part of the review processes of the Genocide Convention which attempted to forge a legal link between the prevention and punishment of ecological crimes, as well as the kinds of cultural genocide that are frequently imbedded in ecological damage (Lindgren 534). However, the proposal has never been implemented and the sole mention of environmental crimes in the Rome Statute was in Article 8 (War Crimes) in 1998, which relates to deliberately creating commonplace, long-term, and serious damage to the natural environment in a war setting (535). In a similar manner, The International Law Commission (ILC) has been engaged in several efforts to enact severe laws of crime against the environment but has failed to do so. All of these failures and the general disregard towards the issue of constructing a universal law against ecocide further leads to more and more ecocide cases.

While environmentalists still find themselves in a struggle to obtain recognition for ecocide as an international crime, cases of ecocide continue to emerge throughout the globe. This section will focus on some of the recent cases of ecocide throughout the globe. *The world's lungs* within its borders, Brazil is a country whose environment is crucial not only to its surrounding countries, but also to the entire world. According to Brazil's National Institute for Space Research (INEP) in 2019, deforestation reached its greatest level in almost a decade, increasing 30 percent from 2018 to 9,762 square kilometers (Raftopoulos and Morley 16). In addition, in 2019, the organization issued a warning of a serious rise in the rate of deforestations across the country and a projected 125,000 hectares of deforested land were burned between July and September of that year, with over 2,500 fires raging in the Amazon by the end of August 2019 (16). Scholars and environmentalists hold Brazil's government and its former head of state, Jair Bolsonaro's policies responsible for the recent ecocide in the Amazon region (18). Furthermore, Bolsonaro has concentrated on promoting the misleading impression that protected areas and indigenous people pose a substantial danger to Brazil's "development" and Amazon sovereignty where on several occasions, he has made it clear that the Amazon is now "open for business," openly declaring that the rainforests will be subjected to entrepreneurs and investors of all sorts (11). The deforestation of the rainforests poses a serious threat to the indigenous population of the Amazon whose population are estimated to be around one million and who are highly dependent on the land (2). In short, ecocide in the Amazon area is sure to have catastrophic effects on Earth as well as indigenous peoples and therefore, must be halted at all costs.

Another case where ecocide has been visible is the Israeli occupied territory of Palestine. The Israeli occupation, and the separation wall in particular, have played and continue to play a significant role in the degeneration of agrarian and grasslands, habitat destruction, biodiversity loss, loss of water resources, and degradation of water quality in Palestine, with drastic social, economic, and political consequences (Short 81). An example related to this is the extreme amount of garbage dumped by Israeli trucks in the occupied areas (82–83). Another consequence of the Israeli occupation is the immense deforestation which was reported that Israeli occupation authorities destroyed 1.5 million trees between 2000 and 2007 in the West Bank and Gaza Strip (83). This scale of deforestation will have disastrous results on climate change, lifestyle and the environment in general. In addition, improper agricultural control mechanisms imposed on Palestinians in the occupied territories have been regarded as a *long-term ecocide* such as water shortages, unregulated home and industrial disposal grounds, and high fertilizer and pesticide use . . . [are] all contributing to in-situ soil degradation (Short 84). To sum up, ecocidal tendencies in Israel-Palestine are known to be at worrying level and need to be dealt with not only for the environment, but also for the well-being of both peoples. However, as usual, it is the will of the strongest that prevails rather than the will of the most righteous.

A third case where ecocide has been experienced repeatedly is Australia. The operation of large coal seam gas seams (CSG) close to the surface has led to gas leakages and grave environmental destruction (Short 147). In the meantime, the Aboriginal peoples of Australia have been left out of the decision-making and implementation process as some support CSG and others do not approve (148). For many Aboriginals the CSG stands for a contradictory selection between colonial manufacture and their native lands but data shows that Aboriginals have little access to CSG projects and their economic benefits (149). Therefore, many Aboriginal groups have created resistance against the CSG. On a tour of Australia's gas areas, *End Ecocide* advocate Polly Higgins stated: "The stories I heard over the last two weeks about CSG, the fracking I saw, and the extreme levels of community concern I experienced led to the question: is this not an Ecocide? Surely it cannot be right to subject our people and planet to gasfield processes that cause significant harm" (Short 151–152). Higgins expressed the sentiment that many natives, scholars and environmentalists alike uphold. As it is the case with the Israeli-Palestinian conflict, those who hold power impose their agenda on those who are subjected to them. Thus, it is imperative that the Aboriginals have a say in all forms of decision-making concerning their soil in Australia.

Ecocide is also observable in Africa where multinational oil companies are inflicting harsh damage on the natural environment. The Niger Delta was a victim of oil pollution caused by these oil firms (Goodnews Loanyie 21). The influence of hydrocarbon chemicals emitted by oil companies on the soil has altered the economic health of the people on a massive scale. Eating polluted agricultural food produced in oil-producing regions has also led to a variety of oil-related illnesses that have wreaked havoc on the people's financial and economic well-being, leading to an increase in healthcare expenses (21–22). Moreover, the question of whether the ecological sustainability marketing activities of the oil firms have amended the economic condition of the population around the Niger Delta can be answered as very low as can be derived from Loanyie's findings which conclude that "agricultural food poisoning, economic poverty, agricultural economic frustration, and agricultural land limitation are caused by oil pollution in the oil-bearing communities in Niger Delta" (44). As a result, it is not surprising to observe multinational capital holders going hand in hand with ecocidal activities. Once profit is involved, ecocide is simply considered collateral damage.

Given these examples, it is apparent that the list of ecocide goes on and on. Many more examples could be delivered from Canada to India, China, Sri Lanka, Russia, Türkiye and many others. However, one case that deserves special attention is the ongoing ecocide in Ukraine. A recent research article published by Tetiana Gardashuk purports that grave and extensive environmental, climatic, natural ecosystems, biodiversity, and habitat destruction becomes visible in Ukraine (3). It has been estimated that around 11 percent of Ukraine's population was economically dependent on agricultural activities which, under the Russian aggression were heavily disrupted and faced serious destruction that could be interpreted as ecocide (Gardashuk 3). This ecocide led to the blockage of 4.5 million tons of grain in Ukrainian ports, disrupting grain export to various parts of the world which will have an immensely negative effect on the increase of famine in African and Asian countries. Gardashuk concludes by affirming that crimes against nature include not only legal but also social, ethical, and existential elements; second, acknowledgment of ecocide crimes belongs not only to the realm of law but also to studies in anthropology, sociology, social psychology, public opinion, advocacy, and activism (4). Hence, as one of the recent atrocities committed in Ukraine, ecocide is sure to have a devastating impact not only on the surrounding region but also on the world as a whole. Following Ukraine, the conflict which evolved into a grave war is the recent ecocide in Gaza that began on October 7, 2023. Since the first days of the conflict, Israeli forces have targeted environmental resources by destroying essential infrastructure with fatal environmental consequences (Shuker). The bombing campaign on Gaza exacerbates a far longer-term problem of Palestinian land loss, which has its own environmental consequences. The relation between water, land and population well-being is intrinsically connected to land acquisition, collective landscape and ecological change (Shuker). It is obvious that this war is an environmental catastrophe, an ecocide presumably worse than that of Ukraine, which will prove to be disastrous for the region and for all the peoples inhabiting the area.

Ecocide is not a national issue but a phenomenon that has serious international impact. It represents humanity's toll on Earth. Frontiers are human made, thus artificial. For Earth, these frontiers do not exist and therefore what takes place in one region has direct influence in the other, regardless of frontiers. For this reason, ecocide has become a vicious circle, a grave problem bound to repeat itself unless a permanent solution is found. All fauna and flora on Earth are living entities, therefore ecocide should not be distinguished from genocide as both include systematic and widespread extermination of living beings. We can no longer afford to be oblivious to the ongoing damage delivered by ecocidal activities. Essentially, human beings ought to build a moral conscience in favour of the preservation of all living beings, including the environment. Sadly enough, it seems that humanity has a long way to go to reach such a level of conscience. Warnings issued by scientists are becoming graver by the day. It is now widely acknowledged that a growing number of environmental catastrophes have their roots in excessive consumption and the ensuing depletion of natural resources (Sarliève 234). Due to all these factors, the only effective and permanent solution to stop ecocide is through the legislation of an international ecocide law. In a recent article published in 2022, Giovanni Chiarini proposed additional amendments to the *Stop Ecocide Foundation* proposal to consolidate the earlier foundation and to include ecocide and aggravated ecocide in the list of special crimes using procedural checks and balances (27–28). Chiarini's suggestions provide vital contributions to the cause.

Furthermore, the eco-conceptualization of ecocide is crystal clear as it represents one of the most crucial issues which requires immediate attention. As a matter of fact, it is one of the most essential eco-concepts that ought to be prioritized. To this end, ecocide as an eco-concept and as a fact ought to be more and more visible on the media, social-media, literature and cinema so as to be able to raise public awareness for the masses around the globe. Scholars and intellectuals are surely aware of it but to create an impact, awareness must be raised in the masses. Therefore, the eco-concept of ecocide must become widespread worldwide, especially in economically struggling and highly populated countries.

In conclusion, many online petitions have been initiated in favour of the ecocide law, some national and others on international scale. I believe it is a moral imperative to provide support for activists of international ecocide law. Therefore, I urge everyone to support and promote the web pages and platforms of the most active online ecocide campaigns at the end of this chapter. In order to obtain result, pressure must be exerted on international authorities, in particular the EU, the UN and the International Criminal Court in The Hague (ICC) to acknowledge and legislate ecocide as an international crime against humanity. So far, the ICC recognizes four types of international crimes which are genocide, war crimes, crimes against humanity, and aggression. Ecocide must be added to the list as the fifth international crime which ought to be fully acknowledged and prosecuted, when necessary. The 1st International Conference on Ecocide was held on November 3–4, 2022 in Istanbul, Türkiye. This promising conference brought together various scholars, activists and environmentalists from different parts of the globe who debated on the alarming issue of ecocide. As a fruitful gathering, I share the aspiration that such meetings will contribute to the cause and therefore ought to be repeated in various locations. All in all, the inability to ensure ecocide as an international crime since the 1970s proves failure and is largely due to the Eurocentric/anthropocentric outlook on the world that remains prevalent, and which must evolve towards an ecocentric view of the world (Dècle-Classen 21). The battle for ecocide is a strenuous struggle which, despite all hardship, must be carried on and brought to completion. Nevertheless, substantial progress has been made on November 2023 proclaiming the passing of new laws by the EU for the prevention of environmental crimes and punishing cases comparable to ecocide. I sincerely hope that this momentous decision will lead to a series of events that will result in the legislation of an effective and permanent international ecocide law.

ECOCIDE LAW ONLINE PLATFORMS & CAMPAIGNS

https://ecocidelaw.com
https://www.ecocidelawalliance.org/
https://www.internationalcrimesdatabase.org/
https://www.stopecocide.earth
https://www.facebook.com/EcocideLaw/
https://www.instagram.com/ecocidelaw
https://twitter.com/EcocideLaw
https://www.youtube.com/c/StopEcocideInternational
https://www.youtube.com/watch?v=HMGw1Y8AmT8 (Republic of Panama's Vice President Addresses the UN General Assembly, September 20–26, 2022)
https://www.youtube.com/watch?v=djNJufsrmDU (Republic of Vanuatu's President Addresses the UN General Assembly, September 23, 2022)
https://youtu.be/PHqIwAkiEPo (1st International Conference on Ecocide, March 11, 2022)
https://youtu.be/Yo0-bSkiZ0U (1st International Conference on Ecocide, April 11, 2022)
https://www.change.org/p/support-making-ecocide-an-international-crime

REFERENCES

Broswimmer, Franz. *Ecocide: A Short History of the Mass Extinction of Species.* Pluto Press, 2002.

Chiarini, Giovanni. "Ecocide: From the Vietnam War to International Criminal Jurisdiction? Procedural Issues In-Between Environmental Science, Climate Change, and Law." *Cork Online Law Review*, 2022. https://ssrn.com/abstract=4072727.

Cullinan, Cormac. "Earth Jurisprudence: From Colonization to Participation." *State of the World, 2010, Transforming Cultures.* The Worldwatch Institute, 2010.

Dècle-Classen, Clémentine. "Ecocide and the End of the Anthropocene: An Ecocentric Critique of the (Failed) Developments of an International Crime of Ecocide." *ICD Briefs 2021/2022*, 2022, pp. 1–21.

Dunlap, Alexander, and Andrea Brock. "Introduction: Enforcing Ecological Destruction." *Enforcing Ecocide: Power, Policing & Planetary Militarization.* Springer Nature, 2022.

Gardashuk, Tetiana. "Is Russian Aggression in Ukraine Ecocide?" *Envigogika*, vol. 17, no. 1, 2022, pp. 1–6.

Goodnews Loanyie, Komene. "Ecocide Activities and Oil Firms' Ecological Marketing Practice: A Focus on Agricultural-Economic Sustainability in Niger Delta." *World Journal of Business and Management*, vol. 7, no. 2, 2021, pp. 18–48.

Govorushko, Sergey M. *Natural Processes and Human Impacts: Interactions between Humanity and the Environment*. Springer Science & Business Media, 2011.

Greene, Anastacia. "The Campaign to Make Ecocide an International Crime: Quixotic Quest or Moral Imperative?" *Fordham Environmental Law Review*, vol. 30, no. 3, 2019, pp. 1–48.

Higgins, Polly. *Dare to Be Great: Unlock Your Power to Create a Better World.* Flint, 2020.

———. *Earth Is Our Business: Changing the Rules of the Game*. Shepheard-Walwyn, 2012.

Higgins, Polly, et al. "Protecting the Planet: A Proposal for a Law of Ecocide." *Crime, Law and Social Change*, vol. 59, no. 3, 2013, pp. 251–266.

Hough, Peter. "Trying to End the War on the World: The Campaign to Proscribe Military Ecocide." *Global Security: Health, Science and Policy*, vol. 1, no. 1, 2016, pp. 10–22.

Lindgren, Tim. "Ecocide, Genocide and the Disregard of Alternative Life-Systems." *The International Journal of Human Rights*, vol. 22, no. 4, 2017, pp. 525–549.

Raftopoulos, Malayna, and Joanna Morley. "Ecocide in the Amazon: The Contested Politics of Environmental Rights in Brazil." *The International Journal of Human Rights*, vol. 24, no. 10, 2020, pp. 1616–1641.

Sangster, Alan J. *Warming to Ecocide: A Thermodynamic Diagnosis*. Springer Science & Business Media, 2011.

Sarliève, Maud. "Ecocide: Past, Present, and Future Challenges." *Life on Land, Encyclopedia of the UN Sustainable Development Goals*, edited by Walter L. Filho. Springer Nature Switzerland AG, 2021, pp. 233–243.

Selwyn, Daniel. "Global Britain and London's Mega-Mining Corporations: Colonial Ecocide, Extractive Zones, and Frontiers of Martial Mining." *Enforcing Ecocide: Power, Policing & Planetary Militarization*, edited by Alexander Dunlap and Andrea Brock. Springer Nature, 2022.

Short, Damien. *Redefining Genocide: Settler Colonialism, Social Death and Ecocide*. Zed Books, 2016.

Shuker, Zeinab. "War Has Poisoned Gaza's Land and Water. Peace Will Require Environmental Justice." *The Century Foundation*, Commentary Century International, December 19, 2023, tcf.org/content/commentary/war-has-poisoned-gazas-land-and-water-peace-will-require-environmental-justice/. Accessed December 24, 2023.

Vorster, Jakobus M. "Ecocide, the Integrity of Creation and the Rights of the Next Generation." *Verbum et Ecclesia*, vol. 26, no. 3, 2005, pp. 867–890.

Wilson, Bruce G., and National Archives of Canada. *Colonial Identities: Canada from 1760 to 1815*. Minister of Supply and Services Canada, 1988.

Chapter 2

Petroculture

Exploring the Limits of Oil, Utopia and Imagination

İsmail Serdar Altaç

Of all the substances that have shaped human history, only a few have enjoyed the privilege that petroleum has had, so much so that a network of culture which now permeates people's lives at a planetary scale has come to be called by the names "oil culture" and "petroculture." Simply put, "petroculture" refers to a set of practices, values, ways of life, and perceptions of the world heavily shaped by the extensive use of oil. Petroculture stands out as one of the most significant eco-concepts, for it crystallizes the hitherto unexplored realm shaped by the interrelations among culture, nature, and energy. Though the materiality of oil in the mere physical operation of modern societies has never been disputed since its modern discovery in the nineteenth century; oil, and energy at large, had largely remained a neutral field concerning culture for the scholars of humanities for a long time. The recent scholarship within the larger field of Energy Humanities, however, unearthed the value-laden and formative role of oil in contemporary culture in manifold ways. Stephanie LeMenager states that "energy systems are shot through with largely unexamined cultural values, with ethical and ecological consequences" (4). The perceived significance of oil has led to a proliferation of neologisms, amongst many others "petromodernity," "petrotopia," and "petropolitics," by which we try to interpret and understand our contemporary existence and experience. However, the crucial role of petroleum and its addictive characteristics have generated certain discourses based on its future, because it is a non-renewable type of energy. This study undertakes a critical engagement with these divergent discourses shaped by the idea of an end to petroculture, as they are shown by Imre Szeman, and lastly attempts to explore a hybrid discourse among them, departing from the notion of "critical utopia."

Though the relationship between energy and culture has been a growing and pressing topic for the last decades, this issue was introduced into social sciences much earlier by the anthropologist Leslie White in his article "Energy and the Evolution of Culture" (1943). His main thesis is that the advancement of human culture is proportional to the amount of energy at his disposal, and the transition into each successive energy system has determined the social structures so far. However, this paper does not delve into an in-depth projection of a post-petroleum society though it hypothetically implies its end in the future: "Should, however, the amount of energy that we are able to harness diminish materially, then culture would cease to advance or even recede" (350). White does not name any specific source of energy in this statement; however, the kind of social system in which he lived had mainly been shaped by the extensive use of fossil fuels. White is cautious about not falling into the pitfall of technological utopianism's naive view that a miraculous solution will be found for any energy shortage even if the existing energy sources deplete. However, published prior to the 1980s when the terms "global warming" and "climate change" were introduced, his study misses the planetary impact of the energy regimes. With its review of the historical continuum, White's argument is well grounded in that it establishes a concrete relation between the use of energy and humanity's development. Surely, the ethical implications of energy regimes are not within the scope of White's article and the lack of this concern does not nullify the validity of his argument. However, viewed from today's vantage point, the study mostly detaches humanity's development from the non-human world on which it is deeply grounded. Imre Szeman underlines the urgency to include "the material resources we have added to the societies" in the discussions of political projections, but he also equally questions the way we use these sources "at a rate and on a scale that threatens the continued existence of humans on the planet" ("Towards a Critical Theory" 26). Nature and culture are, as Arne Naess repeatedly argues, inextricably related: "Humans are clearly inside the ecological systems of the Earth, and . . . [p]rotection of human cultural diversity is a genuine part of the protection of biodiversity" (146).

Even if this relatedness is ignored, another danger lurks in the fact that the current social, political, and economic order is heavily shaped by oil. Imre Szeman calls it "oil ontology" which is "the structuring 'Real' of our contemporary sociopolitical imaginary" ("The Cultural Politics" 34). The question of whether the age of oil is a historical anomaly whose end is likely to bring about even deeper anomalies or just a step in the progressive march of civilization is impossible to answer. Frederick Buell aptly calls the current period "'an age of exuberance' . . . haunted by catastrophe" (71). The comforts provided by this fluid, along with the augmented power it confers, approximates our experience of it to that of Icarus. The "maze" of old hierarchical structures tethering one's social mobility was overcome by the pervasive use of oil, to a certain extent, as the conditions created by "fossil fuels opened up many new avenues for the accumulation of wealth" (Bergthaller 428). Thus, as shown in the exemplar *Great Gatsby*, it becomes one of the imageries associated with upward mobility and self-fashioning. Having said that, just like Icarus' fate, which curiously resembles Hubbert's peak, everything enabled by oil lingers on the verge of "catastrophe." The opacity of a future determined by oil has become a source of inspiration for diverse discourses.

Oil is a strange fluid in cultural imagination in that the scenarios of its absence produce more diverse and intriguing narratives than its current existence. Out of a wide array of narratives, Imre Szeman detects three fundamental types of interpretation born out of the consciousness of peak oil. These narratives can be broadly classified as "strategic realism," "techno-utopianism," and "eco-apocalypse" ("System Failure" 808). These symbolic forms are ideologically charged and focus on different temporalities. Strategic realism, associated with right-wing politics, can be defined as a (voluntary) temporal myopia in which the governments solely focus on the appropriation of natural energy sources, "suspend[ing] or minimiz[ing] concerns about the cumulative environmental disaster of oil or the fact that oil is disappearing altogether" ("System Failure" 810). This is the discourse that has informed much of the geopolitical conflicts since the early twentieth century. The other type of discourse, techno-utopianism, presupposes either the invention of new technologies that can re-enable the easy extraction of oil which is getting harder each passing day, or a smooth transition into a new energy source when the oil sources are depleted (though there is no concrete evidence of it). At face value, the realization of the former means being a step closer to the brewing climate catastrophe while the latter's lack of evidence may lead to a myopia of a possible social crisis. To re-state it in Gerry Canavan's words, "either we have peak oil and the entire world suffers from a tumultuous transition to post-cheap-oil economics, or else there is plenty of oil left for us to permanently destroy the global climate through excess carbon emissions" (333). However, despite the high level of precision in this argument, it might be slightly reformulated. If the transition predates a climatic catastrophe, the global political and economic order is likely to suffer; however, if these events happen the other way around, the global order will not be immune to the changes taking place in the environment. The perceived difference between the fates of human and non-human owes something to an image of nature as "a surrounding medium" (Morton 4–5). The division between "nature" and "culture" shaped by the logic of modernity obscures the possibilities of change. As Dipesh Chakrabarty indicates human and non-human histories have reached a point of convergence wherein "the warming of the planet threatens . . . not the geological planet itself but the very conditions, both biological and geological, on which the survival of human life as developed in the Holocene period depends" (213). Thus, the quest/wish for re-introducing cheap oil and collective inertia relying too much on the scientific processes whose results have not been yielded are akin to each other as to the social prospects. While the fictional representations speculate on the climate-related displacement of human beings in the near or distant future (De Bruyn), a preview of it is already set in motion in reality. UNHCR reports that the world has witnessed the uprooting of more than 20 million people annually since 2008.

This dream of a "smooth landing" in techno-utopianism shares the complacency of strategic realism about a future disorder. They "operate within the existing understandings of the way the world operates" (Szeman "System Failure" 816). Neither of them has any hesitation about the politics of extraction that have constituted the configurations of capitalism over centuries. Placing hopes on science just because of what it has offered so far lies at the core of techno-utopianism. However, as Dipesh Chakrabarty's inquiry into the origins of how humanity became a geologic force shows, the current mastery over the environment is a result of historical contingency: "Human civilization surely did not begin on the condition that, one day, in history, man would have to shift from wood to coal and from coal to petroleum and gas" (216). Techno-utopianism is at a loss to see this contingency, contradicting the very logic of Enlightenment on which it apparently rests.

Eco-apocalypticism is another discourse founded on the scenarios of the end of oil. This discourse, according to Szeman, "makes it clear that disaster cannot be avoided without fundamental changes to human social life" ("System Failure" 816). The revolutionary spirit that defines eco-apocalypticism mainly criticizes the patterns of consumption that put a greater demand on oil sources. What distinguishes this discourse from the others is that it recognizes the social and environmental limits of the current petroculture. Szeman does not sound optimistic about the praxis of this discourse, espousing that "since such change [social and political] is not on the horizon . . . it sees the future Bosch-like—a hell on earth" ("System Failure" 815). However, due to its "*pedagogic*" (816) aspect, it can be said to have certain level of utopianism. The utopianism of this discourse has a different target than that of technological utopianism. While the former seeks to change society, the latter dreams of changing the energy while keeping the existing social metabolism and habits. Thus, the prospective pessimism of eco-apocalyptic discourse might also mobilize an urge to act here and now, rather than a self-defeating inaction or a blind faith in science. For this reason, it might be appropriate to categorize those intentional communities of ecological sustainability under this discourse. Transition Network is one of the initiatives that seeks to map solutions for a post-oil society and resilience through localized community experimentations and permaculture practices. Rob Hopkins, the founder of the network, defines their basic assumptions as lower energy consumption, introduction of resilience, collective action, and respect for biological limits (134). Though Hopkins defines Transition Network as "an immersion in the possibilities of applied optimism" (15), it is, after all, a product of looming disaster scenarios about the environment and social order if the current growth mentality endures. To state it more succinctly, it has a utopian impulse born out as a reaction to a moribund social order. Scattered across the globe, the branches of the network establish permaculture communities that adopt the principles of decentralization and decarbonization.

However, this discourse is not without its pitfalls, as one might imagine in any initiative that has a utopian context. Szeman is cautious about the optimism of these groups: "there is no guarantee that the new world on the other side of the end of oil will be one made in the images of revolutionary groups and their labors" ("Crude Aesthetics" 439). What G. C. Unruh calls "carbon lock-in" which fossil fuel regimes have engendered leaves little room for a large-scale revolutionary attempt, which raises doubts about transition without social turbulence. For instance, J. H. Kunstler's *A World Made by Hand* offers a portrayal of a well-operating, if not a utopian, post-oil society which, nevertheless, was haunted by a history of decimation of global population by war and epidemics. The crucial question is whether this "world" would be possible if it had not been for the demographic contraction. This does not mean that futuristic communitarian and revolutionary imaginings are doomed to fail just because of an inherent political flaw in their agenda, but that the world has evolved in a way that irrevocably exceeds the limits of the planet. The human carrying capacity of the earth, which has skyrocketed thanks to fossil fuels (Friedemann 22; Buell 71), is likely to shrink back to its pre-fossil fuel levels due to lack of fertilizers (Szeman "System Failure" 816). The previous transitions took place in a way that released human beings from the constraints of nature and its rhythms, thus boosting the global population. However, unless the techno-utopian predictions are realized, the permaculture experimentations of today are unlikely to satisfy the whole population even at subsistence levels. According to the novelist and social critic, J. H. Kunstler, the looming disaster is most likely to revolve around agricultural practice which has already lost its flexibility as a result of such factors as current urban planning, industrial farming, and property rights (17, 241, 244). The challenges faced by this movement are not only limited to food production and distribution systems but also consist of possible predicaments resulting from alternative energy regimes. Despite being a supporter of non-fossil fuels, Vaclav Smil emphasizes the need to acknowledge the insufficiency of renewable energy sources to fully address a post-oil future, calling the transition "desirable," yet "more difficult than [it] is commonly realized" (108). If the objective is the survival of the human species, this may be achieved if the oil sources deplete prior to a climate cataclysm, but if the objective is the survival of all human beings, the current alternative energy regimes and production methods seem to be hardly sufficient to address the problem. However, according to another view, disasters "are opportunities to enhance resilience" (Zebrowski); they need to be embraced for radical social transformation. Unlike the abovementioned optimism of the Transition Network, the anxieties caused by a possible loss of this energy source inform many science fictional and dystopian narratives in which, as Gerry Canavan maps them in "Retrofutures and Petrofutures: Oil, Scarcity and Limit," the scarcity leads to either world state authoritarianism or a conflict over the control of natural resources.

In the literary representations, it might be possible to track another vein that consists of both of the futuristic projections about oil. Departing from Tom Moylan's conceptualization of "critical utopia" in literature, one might capture some utopian texts that deal with the issue of petroleum, be it as the main or a minor topic, as an ongoing problem within the utopian context. Following the liberatory movements set in motion in the 1960s, literary utopianism reconfigures in a way that ends up in the emergence of "critical utopias" which repudiate idealized and rigid alternative worlds of the traditional utopias while retaining its utopian impulse. Evading the pitfalls of the former utopias, critical utopias acknowledge "difference and imperfection within utopian society itself and thus render more recognizable and dynamic alternatives" (Moylan 10). Being one of the prominent examples of critical utopia, Ursula K. Le Guin's *The Dispossessed* (1974) portrays an imperfect utopian society that instantiates the limits to decentralization. Though petroleum is not the driving force, the logic of petroculture is still at work in the novel. Odo's plan for a decentralized political system is, in practice, challenged by Anarresti harsh geography, for "her plans . . . had been based on the generous ground of Urras" (77), and, as a result, a vast network of roads is required for the communities to interchange resources among each other. The interdependence between communities is a natural component of social ecological thought, yet the ensuing requirement of roads may be read as one of those inherent problems bequeathed by the capitalist (A-lo) and state-communist (Thu) order. Though the Anarresti have harnessed sufficient energy through renewable methods, they still rely on the petroleum which they import from Urras, much to the dismay of the Anarresti.

Laurence Manning's *The Man Who Awoke* (1933) provides another telling example of petroleum in a utopian context. Though it was not written under the political circumstances that nourished critical utopia, it still reflects an imperfection within utopia. Norman Winters, the protagonist of the novel, travels in time, only to find a future decentralized society much like Anarresti. As a utopian hallmark, this society does not suffer from a serious physical want and even has a transportation system whose efficiency far exceeds that of the petroleum age. Furthermore, confirming the validity of techno-utopian visions, the society uses a renewable energy source, "wood alcohol," in their aircrafts. Despite the rapport in this society, Winters confronts with a severe criticism of the habits of over-consumption in the twentieth century: "But for what should we thank the humans of three thousand years ago? For exhausting the coal supplies of the world? For leaving us no petroleum for our chemical factories?" (21). In fact, what is at stake here is more of a case of sustainability regarding all resources than a grumbling over the mere absence of petroleum. Nevertheless, their oblivion to the impact of fossil fuels on nature which they hold so dear sounds ironic.

J. G. Ballard's "The Ultimate City" (1976) portrays an even more curious case of petroleum. Garden City, a post-oil settlement "far more sophisticated in every respect than that of" (Ballard 878) the societies depending on fossil fuels, has a smooth functioning social order. Despite the Arcadian perfection, the main character, Halloway, grows dissatisfied with the slow rhythms of life, which eventually urges him to leave the settlement in search of a more vibrant and hectic atmosphere. His transformation is marked by his innate desire for ever more power at his disposal, though he neither enjoyed the amenities of the petroleum age nor witnessed the transition in person. As it were, the craving for more energy is implied to have become almost a phylogenetic human trait. Though the society at large settles for a low-energy regime, an individual eccentricity taints the utopian fabric.

These narratives, whether they take energy as their primary point of interest or not, occupy a liminal space about oil futurities. Respectively, they demonstrate the political, ethical, and psychological dimensions of petroculture in a critical utopian context. While they embody a utopian impulse, they are marked by "the awareness of the limitations of the utopian tradition" (Moylan 10). Though these representations display societies, in many ways more reformed than the current one, they are still unable to move fully beyond the limits of imagination as to the post-peak oil world. Vaclav Smil's position is probably one of those closest to the critical utopian view of oil futurity: "The world beyond oil is still many decades away but we should see the path toward that era as one of very challenging, but also immensely rewarding, opportunities as modern civilization severs its dependence on fossil carbon" (*Oil* 188). On the other hand, techno-utopian and eco-apocalyptic narratives face their own challenges in an age deeply immersed in petroculture. Plato's cave metaphor may somewhat illustrate the limitations faced by each of the narratives discussed in this chapter. Having been acculturated by an energy-intensive lifestyle, the modern subject is incapable of fully overcoming the epistemological barrier between himself/herself and the future of oil (culture). However, utopianism that calls for a radical social change, observed in the eco-apocalyptic discourse and critical utopian narratives, seems to be the only viable option in our attempt to brace for the impact. An understanding of the eco-concept of petroculture may not provide a crystal-clear view of its future but secures a vantage point that enables self-scrutiny for contemporary society in its relation to energy.

REFERENCES

Ballard, James G. *The Complete Short Stories*. W.W. Norton, 2009.

Bergthaller, Hannes. "Fossil Freedoms: The Politics of Emancipation and the End of Oil." *The Routledge Companion to the Environmental Humanities*, edited by Ursula K. Heise, Jon Christensen, Michelle Nieman, Routledge, 2017, pp. 440–48.

Buell, Frederick. "A Short History of Oil Cultures; or, the Marriage of Catastrophe and Exuberance." *Oil Culture*, edited by Ross Barrett and Daniel Worden, University of Minnesota Press, 2012, pp. 69–88.

Canavan, Gerry. "Retrofutures and Petrofutures: Oil, Scarcity and Limit." *Oil Culture*, edited by Ross Barrett and Daniel Worden, University of Minnesota Press, 2012, pp. 331–49.

Chakrabarty, Dipesh. "The Climate of History: Four Theses." *Critical Inquiry,* vol. 35, no. 2, 2009, pp. 197–222.

De Bruyn, Ben. "The Great Displacement: Reading Migration Fiction at the End of the World." *Humanities*, vol. 9, no. 1, 2020, https://doi.org/10.3390/h9010025.

Friedemann, Alice J. *Life After Fossil Fuels: A Reality Check on Alternative Energy.* Springer, 2021.

Hopkins, Rob. *The Transition Handbook: From Oil Dependency to Local Resilience.* Green Books, 2014.

Kunstler, James Howard. *The Long Emergency: Surviving the Converging Catastrophes of the Twenty-First Century.* Atlantic Monthly Press, 2005.

Le Guin, Ursula K. *The Dispossessed.* Avon Books, 1974.

LeMenager, Stephanie. *Living Oil: Petroleum Culture in American Century.* Oxford University Press, 2014.

Manning, Laurence. *The Man Who Awoke.* Ballantine Books, 1975.

Morton, Timothy. *Ecology without Nature.* Harvard University Press, 2007.

Moylan, Tom. *Demand the Impossible: Science Fiction and Utopian Imagination.* Edited by Raffaella Baccolini. Peter Lang, 2014.

Naess, Arne. "Culture and Environment." *Humboldt Journal of Social Relations,* vol. 46, no. 2, 1994, pp. 143–49.

Smil, Vaclav. *Energy Transitions: History, Requirement, Prospects.* Praeger, 2010.

———. *Oil: A Beginner's Guide.* Oneworld, 2008.

Szeman, Imre. "Crude Aesthetics: The Politics of Oil Documentaries." *The Journal of American Studies*, vol. 46, no. 2, 2012, pp. 423–39.

———. "System Failure: Oil, Futurity, and the Anticipation of Disaster." *South Atlantic Quarterly*, vol. 106, no. 4, 2007, pp. 805–23.

———. "The Cultural Politics of Oil." *Polygraph,* vol. 22, 2010, pp. 33–45.

———. "Towards a Critical Theory of Energy." *Energy Humanities. Current State and Future Directions*, edited by Matúš Mišík and Nada Kujundžić, E-book, Springer, 2021, pp. 23–36.

UNHCR. Frequently Asked Questions on Climate Change and Disaster Displacement, 1 December 2023, https://www.unhcr.org/news/stories/frequently-asked-questions-climate-change-and-disaster-displacement.

Unruh, Gregory C. "Understanding Carbon Lock-in." *Energy Policy*, vol. 28, no. 12, 2000, pp. 817–30.

White, Leslie. "Energy and the Evolution of Culture." *American Anthropologist*, vol. 45, no. 3, 1943, pp. 335–56.

Zebrowski, Christopher. "Governing the Network Society: A Biopolitical Critique of Resilience." *Political Perspectives,* vol. 3, no. 1, 2009, https://repository.lboro.ac.uk/articles/journal_contribution/Governing_the_network_society_a_biopolitical_critique_of_resilience/9503633

Chapter 3

Dark Ecology

Embracing Strange Intimacies and Withdrawals in the Joyful Collapse of Nature

Clara Soudan

REOPENING ECOLOGICAL THOUGHT AFTER THE FALL FROM NATURE

As it delves into the possibilities of an ecological thought emancipated from ideologies of Nature, the work of Timothy Morton weaves a redefinition of ecological criticism calling into question some of the most established concepts of ecology. Understandably, the recent consciousness of the planetary magnitude of anthropogenic actions compels a renewed understanding of our being in the world. Drawing from the observation that ecological awareness challenges us to think and feel at different scales, thus disorienting concepts of nature, life, species or even thought, Morton embarks upon a reassessment of philosophy, politics and art in light of the ecological catastrophe. The term dark ecology designates their effort to weave an ecological thought freed from nature and committed to an openness which remains unthought. Illuminating what the concept of nature obscures, dark ecology as an *eco-concept* delves into a world unconquered, which eludes modern ontology and epistemology, thus offering a key contribution to ecocritical thought. Inspired by Derridean hauntology,[1] Morton's philosophy resembles an investigation of the spells we live under: it studies the concept of nature's tendency to haunt our ontological narratives and therefore the way we dwell in the world. Amidst clouds of radioactive waste and the threatening tick of the Doomsday Clock lingers the overarching spectral presence of the concept of nature. Morton asserts that "one of the things that modern society has damaged, along with ecosystems and species and the global climate, is thinking." Claiming that modern thought has established nature "as a reified thing in the distance, under the sidewalk, on the other side where the grass is always greener, preferably in the mountains, in the wild," they expose an alienated and dualistic structure of thought conditioning our ecological consciousness (*Ecological Thought*[2] 3). In *Ecology without Nature*,[3] Morton thus lays the groundwork for an environmental aesthetics open to ways of experiencing, perceiving and inhabiting the world impeded by the narratives of nature. Dissolving boundaries between cultural and natural, uncovering the precariousness and contingency of our porous trajectories of becoming, the ecological apocalypse manifests a radical entanglement with the more-than-humans which the concept of nature fails to approach. Dark ecology, I argue in this chapter, emerges from the collapse of the dualistic order of nature: amidst ecological devastation, in the ruins of traditional dichotomies, it arises as an attempt to recover from conceptual, normative and political structures of alienation from the more-than-human world.

Following in the footsteps of anthropologist Philippe Descola and philosopher Bruno Latour, Morton deconstructs the concept of nature as an ideological artefact, a totalizing abstraction acting in modern cosmology as a "transcendental principle" and conditioning an impoverished engagement with the world. Their refusal to "digest" actual contingent beings into reified ideal entities and thereby petrify the trajectories of their becoming leads them to discard the concept of nature for extinguishing the radical otherness and openness of the world (*EwN* 13, 262). As a modern concept, nature connotes indeed qualities of "hierarchy, authority, harmony, purity, neutrality" unapt to render the radical queerness of our entanglements with the more-than-human and therefore "[failing] to serve ecology well" (*ET* 3). In *Facing Gaia*, Bruno Latour uncovers nature as a political regime resting upon an ideological apparatus which distributes agencies across modern cosmology and along a spurious nature/culture binary. Latour shares the concern for an ecological thought emancipated from the cosmology of nature and from modern narratives of estrangement disregarding our diffuse and complex attachments to the world. As they remind us that the interconnectedness with all things lies at the core of ecological awareness, Morton invites us to step across disciplinary binaries and cultivate unyielding intimacies. Interestingly, both Latour and Morton suggest bypassing the dualistic route of nature rather than attempting to overcome it, advising that "instead of 'getting over nature,' perhaps we need to get under it, or go through it" (*Dark Ecology*[4] 255). Doing so, they offer to slow down—withdrawing at least temporarily from the race toward successive paradigmatic shifts which Thomas Kuhn had identified at the core of modern scientific practices—to instead dive deep into the intricate paradoxes of nature. Conscious that our thoughts are deeply rooted in a conceptual ecosystem and that paradigms are "viscous" (Latour, *Chroniques* 28), driven by the intuition that something in our relationship to the world hides away and remains unexhausted by the ideology of nature, dark ecology reclaims the obscurity unveiled with the fall of dualism. It embraces a movement of staying with the trouble of a cosmological collapse, dwelling in an opaque cloud of unknowing and diving deeper into the mysterious darkness of our earthly entanglements. Morton observes that "environmental awareness might have something uncanny about it, as if we realized we were caught in something." As the fall from nature unearths an inescapable entanglement, an intimate implication with the otherness of the world, "the collapse of distance dissolves barriers separating subject, object and abject" (*EwN* 58, 215). From the ruins of nature, unveiled as "inhuman, radically different, irreducibly strange, threatening to our need for coherence," a swarming world of strange intimacies emerges through a dazzling cloud of unknowing (*DE* 265). To approach such a cloud opaque and porous, to think of the radical interdependence that binds us to the world while acknowledging both the intimacy and the distance, the holes and the threading, Morton recourses to the metaphor of the mesh. "Radically open," without centre or edge, the mesh provides a figure of deep interwovenness that clears a space for what withdraws itself from our perception ("The Mesh" 22).

Woven throughout Timothy Morton's writings, dark ecology delineates an ecocritical thought as well as an ecological consciousness released from the reductions and commodifications of nature. A key concern of dark ecology thus lies in reclaiming the "radical openness of ecological thought" which has been obscured by a "totalizing idea of nature": "like a dam, Nature contained thinking for a while, but in the current historical situation, *thinking is about to spill over the edge*" (*ET* 11, 3, my emphasis). How does the present ecological predicament make thinking overflow its established categories? What does the radical openness required by ecological thought disclose that the concept of nature conceals? Committed to that which withdraws itself—or alternatively overflows—from the construct of nature, Morton summons us to think through the ecological urgency beyond sterile conceptual and disciplinary compartments. Doing so, they recall ecology to its most radical definition: "thinking how all beings are interconnected, in as deep a way as possible" (*DE* 255). Dark ecology offers ecological criticism an alternative to nature: one that acknowledges and embraces the inescapable paradoxes of our entanglement with the world. Dark ecology thus calls upon us to embrace a radical intimacy with radical strangers heralded by the renewed realization of our planetary entanglement. Morton argues that ecology demands an intimacy with other beings which the concept of nature fails to deliver, namely an aesthetic sensitivity to the presence of that which is intrinsically withdrawing—unsubsumable others relentlessly permeating our own being and yet escaping all attempts to grasp them. Just like the strange intimacies we weave with other beings of the world, human and more-than-human, ecological thought is radically complex and queer (*EwN* 247). Embodied in what Isabelle Stengers names the "intrusion of Gaia," the consciousness of our radical and inextricable involvement within earthly processes of becoming "threatens our need for coherence." Combined with the vertiginous, geological magnitude of our interdependent actions, it shakes up the gratifying, "comfortable way in which humans appear in the foreground and everything else is in the background" (*DE* 256, 265). Dark ecology thus offers to think what this intrusion of a "forgotten form of transcendence," ticklish and indifferent, breaks open in terms of trajectories of thought as well as concrete becomings (Stengers 47).

"WEIRDNESS, THE ELIXIR OF LIFE": SHADES OF ECOLOGICAL DARKNESS

Emerging ecological awareness may occur as a gloomy "turn of event," a tragic discovery of the "unexpected fallout from the myth of progress" or from global capitalism (*DE* 7). Understandably, the prospect of unretrievable environmental loss—of species, habitats, life-forms—may induce overwhelming feelings of grief, depression, despair or anxiety. Earth system sciences tell us indeed about catastrophic trajectories of mass extinction, ecosystem collapse, nuclear radiation, apocalyptical climate change. For Morton, tuning in to the present ecological reality is necessarily "weird" and implies a feeling of uncanniness that should be embraced. "Isn't this lingering with something painful, disgusting, grief-striking, exactly what we need right now, ecologically speaking?" (*EwN* 264). Morton notes that the uncanny dimension of environmental awareness pertains to a feeling of being "caught in something" (*ET* 58)—maybe some strange and intimate mesh? Or the multi-species grip of Chthulu? Reminiscent of Donna Haraway's invitation to "stay with the trouble" of our tentacular entanglement on a damaged earth, dark ecology summons us to linger in darkness, and "stay right here, in the poisoned mud," from which there is no escape. With Joanna Macy, Morton suggests we stay at least some time with the grief of an absolute, radical loss that entails no possible sense of closure (*EwN* 269, 252). In *Landing on Earth*, Latour similarly reminds us that—despite certain hubristic dreams of modernity—there is no escaping our earthbound condition and the inextricable entanglements that come with it. But ecological darkness is manifold: shot through with uncanniness, it begins in depression, then goes through ontological mystery, to end as sweetness. One of its variations presents itself as vertigo.

According to Morton, the event which baptized a new geological era—thereby reawakening us to the planetary magnitude of our entanglements—only appears as the toxic unfolding of a concept of nature "waiting to emerge as catastrophe" (*DE* 59). The Anthropocene locates us at the intersection of human temporality and geological time, each individually part of a geological entity on a planetary scale. Blurring boundaries of time, species, organism or environment, the consciousness of a new geological era defined by the influence of anthropogenic actions on the scale of Earth history induces a vertigo at once existential and cosmological. Various proposals for alternative terms such as Capitalocene (Malm), Chtulucene (Haraway), Planthropocene (Myers) or Androcene (Bahaffou and Gorecki) have pointed out the political limitations of the Anthropocene narrative, including its failure to address some cardinal discrepancies within what is depicted as a unified human condition. The Anthropocene fails indeed to account for unequal responsibilities as well as structures of power, domination and exploitation among humans. Yet Morton, who overall refrains from a critical distance towards the universalist and homogenizing tendencies of the Anthropocene[5], acknowledges the legitimacy of a framework which challenges us to think at earth's magnitude and become aware of our ontological intertwinement with earthly processes. Morton thereby focuses on the phenomenological, almost existential event of the Anthropocene, at the risk of neglecting its embodied, contingent, eminently political dimension. They approach the ungraspable idea of a human species as a "hyperobject," characterizing objects so massively distributed in time and space as to transcend and confound traditional self-oriented frameworks. Like plutonium radiation, Styrofoam or climate change, the human footprint on earth is now irreversible, forever existing on "almost unthinkable timescales" (*ET* 19). Dark ecology thus registers the vertigo that comes with the task of thinking at temporal and spatial scales that are unfamiliar, even "monstrously gigantic," and with an existence lived on simultaneous scales, individual and planetary. If the notion of belonging to a species appears "open, porous, flickering, distant from what is given to perception," the ecological processes we are involved in require us to embrace such a confounding scale (*DE* 24). In this regard, as Latour also demonstrates in *Landing on Earth*, the ecological predicament introduces a sense of radical disorientation, on a political as well as on an existential level: illuminating our place in the biosphere, it imposes upon us a change of perspective and—hopefully—of trajectory. This is notably illustrated by the tenacity of global warming negationists and other reactionary forces of modernity who fiercely oppose the radical openness—at once historical, political and intellectual—unearthed and called for by the ecological crisis. As historian Jean-Baptiste Fressoz reminds us, the ecological apocalypse has long crystalized political controversies around the possible trajectories of technoscience, progress and modernization. In the *Happy Apocalypse*, he recalls the political and intellectual forces which "led us on the path to the abyss." Fressoz thus describes a centuries-long process of disinhibition to the dangers of what came to be broadly accepted as technological progress. The ignorance-producing devices of this "modernizing unconsciousness" have struggled to repress and occult a reflexivity that was always already present (Fressoz 12, 9). But such reflexivity seems inescapable. Morton describes this quality of ecological awareness, which reaches us like a vertigo as the planetary boomerang (un)expectedly returns, in terms of loops.

The "loop-form" of ecological phenomena captures their radical interaction, their interconnectedness as well as their intrinsic sensitivity to each of our actions: "ecological reality consists of porous boundaries and interlinked loops" (Morton, "Why Ecological Awareness is Loopy" 106). Echoing Galileo's discovery that "the earth moves," the philosopher Michel Serres suggested that we were now witnessing another trial, where earth system scientists would claim that "the earth is moved" (1990). Awakening us to the loop-form of things, the current ecological catastrophe amounts to a revolution not unlike the discovery of heliocentrism: the discovery that the earth is moved, touched by us, and that a ticklish assemblage of multi-species processes always reacts to what we put into the world. *The earth is responding*, it retroacts, shivers, overflows under our feet—how unfathomable are their Gaian ways! This planetary responsiveness contributes for Morton to the radical weirdness of ecological awareness, which ultimately unveils the loop-form of all things: human interference, biological systems—everything exists in relation to one another, *everything returns and responds*. The irruption of Gaia, or the sudden consciousness of the presence of more-than-human agencies in the world, might be perceived as an interference, a disturbance to the modernizing logics of extractivism and productivism. Ecological awareness implies an openness to the unexpected outcomes of entangled processes, to the obscure feedback loops of interactions between humans and more-than-humans. Not only does it require a response-ability on our human end, but it also acknowledges a non-human right to respond, in their own cryptic languages, to the assaults they experienced. But this "dark pathway" into the bewildering loop-form of things, Morton argues, has been obstructed, at times even suppressed by dominant Western philosophy.

Morton identifies *agrilogistical thought* as a ten-thousand-year process of trying to rid the world of its weirdness and "un-loop the loop form of things." What they term "agrilogistics" or "the machinery of agriculture" refers to a viral and widely successful approach to human-built space which arose in the Fertile Crescent as an answer to the human desire to eliminate anxiety, notably concerning the access to food security. They write: "agrilogistics promises to eliminate fear, anxiety, and contradiction—social, physical, and ontological—*by establishing thin rigid boundaries between human and nonhuman worlds* and by reducing existence to sheer quantity" (my emphasis). Demarcating human systems from Earth systems, this agricultural logistics lies at the roots of the concept of nature and its connate nature-culture split, which allows the systemic repression of a paradoxical realm of human-nonhuman relations. For thousands of years, eventually requiring steam engines and industry to endure, agrilogistics has implemented the dubious delineation between agricultural and natural, thereby "establishing the necessarily violent and arbitrary difference between itself and what it 'conquers' or delimits" (*DE* 109, 43, 43). Echoing a key claim of ecofeminism, Morton suggests that all dynamics of oppression and exploitation rest upon a fundamental caste structure originally distinguishing between humans and more-than humans.[6]

Embracing what the agrilogistical thought has repressed, dark ecology delves into the primordial relatedness of humans and more-than-humans. It explores the modalities of an ecological thought emancipated from agrilogistical logic and its culture of alienation from the more-than-human world: "for it is my contention that this temporality is very much with us right now and that it provides a way to think how to unwind the catastrophe of agrilogistics." Morton does not resist the temptation to coin another term to describe what precedes, resists and overflows such agrilogistical culture: *the archelithic* thus refers to "a possibility space that flickers continually within, around, beneath and to the side of the periods we have artificially demarcated as Neolithic and Palaeolithic." In this realm of possible, "the magical, flickering aspect" of reality infuses an irreducible sense of uncertainty. "Archelithic cultures of magic" always include what Morton calls "the possibility of pretense"—the possibility of being deceived and overcome by a mysterious cloud of unknowing. Espousing the "dance between concealing and revealing," they intertwine skepticism and faith without opposing them. Open to wonder, embracing all the weirdness of the world, unafraid of the rift between what things are and what they appear, dark ecology lingers in between, clearing a space for what withdraws from our perception and understanding. Acknowledging the opacity of our earthly entanglements, it opposes the proliferating logic of agrilogistics, which "proceeds without stepping back and rethinking" (*DE* 77, 81, 85, 42). Dark ecology thus discloses the preservation of the ontological and epistemological uncanniness of the world as a key issue of ecology. Ecological thought intrinsically reflects upon something overwhelming: an absent presence, an intangible scope, and the strange intimacies binding us to the other beings of the world.

ECOLOGY AS AN ART OF STRANGE INTIMACY: TOWARD AN APOPHATIC THOUGHT OF OUR DARK ENTANGLEMENTS

Morton's critique of the concept of nature illuminates its failure to register a particular kind of intimacy attaching us to the more-than-human world. In "the Dark Ecology of Elegy," Morton reminds us that "ecology is profoundly about intimacy," having to do with an uncanny intricacy with other beings—humans and more-than-humans alike (257). Uncovering intimacy as a fundamental category of ecological thought, dark ecological thought delves into the obscure entanglements attaching us to the world and embraces the radical dimension of a withdrawing, eluding attachment. It dwells in this opaque liminality between beings. "Saturated with unrequited longing," it roams the nebulous space between strangers radically and intricately entangled in each other, exploring a strange intimacy that haunts our relationships with other beings of the world (*EwN* 186). An "encounter with the radically unknown," unearthing a "troubled and troubling intimacy" with radical strangers, ecological awareness is thus illuminated as an experience of the uncanny (Elegy 257, 253). It manifests our being "profoundly covered in, surrounded by and permeated by all kinds of entities that are not (us)" ("What Is Dark Ecology?" 56). The prism of dark ecology thus unveils the ecological issue as one of unfathomable intimacy with the strange, (re)awakening us to a profound wonderment about the presence of others within us—one that should be preserved and nurtured rather than subsumed or repressed. Such an intimacy is inextricable, summoning us to coexist with a host of strangers surrounding and penetrating us—crawling, walking, flying, visible and invisible people of the world. For Morton therefore, "ecology is about relating not to Nature but to aliens and ghosts": it discloses an intimacy full of "ambiguity and darkness." Their concept of *strange strangers* articulates the uncanny entanglement manifested in ecological awareness. Dissolving boundaries between inside and outside, self and other, the strange stranger confounds our sense of self to unveil an intimacy with all things: "since there is no (solid, lasting, independent, single) self, we are the strange stranger." This intimate stranger is inexhaustible: heralded in a multitude of life-forms entangled in another, it becomes weirder as you get closer. As it embraces the ambiguities of a radical intimacy with radical strangers, dark ecology erodes the traditional concept of self to espouse a thought of radical entanglement: "nothing exists by itself, and so nothing is fully 'itself'" (*ET* 100, 87, 14). In this respect, the strange stranger "is itself, yet uncannily not itself at the same time" (2017 18).

Describing the disorienting strangeness of our intimacy with other beings, Morton raises the issue of how to care for and respond to our entangled neighbors as a key aporia of ecological ethics: "unable fully to introject or digest the idea of the other, we are caught in its headlights, suspended in the possibility of acting without being able to act" (*EwN* 249). Dark ecology unfolds in the consciousness that we are entangled with unsubsumable others, our intimate strangers, in processes of becoming full of contingencies, artificialities and hybridities. How to act from within this radical entanglement? Morton answers: "the most ethical act is to love the other precisely in their artificiality, rather than seeking their naturalness and authenticity." Dark ecological thought is indeed concerned with avoiding the petrification, fetishization and reification of others in alienating abstractions which tend to reduce the complex processes of hybridisation constitutive of our entangled becomings. Against ideologies of nature, dark ecology "refuses to digest plants and animals and humans into ideal forms" (Elegy 269). Rather, it is committed to preserving the radical otherness of strange strangers along with the openness of their becoming. Devoted to a cloud of intimacy and distance surrounding our entanglements, dark ecology pledges to clear a space for what manifests and yet withdraws itself from our grasp. It thereby echoes Martin Heidegger's withdrawal of Being, which approaches the unfolding of Being in the light of an active resistance in its manifestation—an inherent elusiveness, a presence that discloses itself through its very concealment. Morton's philosophy of dark ecology embraces—even celebrates—this profound ungraspability of things, so near and yet perpetually withdrawing, as a radical dimension of our being intricately entangled within the otherness of the world. Calling us to get closer to this withdrawing presence, it waves toward Luce Irigaray's philosophy of alterity and her idea of nearness: the feminist philosopher depicts indeed an otherness intimately woven through the self—"the other is already within her"—an alterity "so near that she cannot have it, nor have herself" (Irigaray 31). Abiding in the liminal realm of the intimate, right in between the blurred boundaries of self and other, beings, species, organism or environment, dark ecology invites us to delve further into this liminality and unfold a reflection around the queerness of our entanglements. It calls upon us, in David Abram's words, to tend the porous boundary between beings, to dwell on the edge of worlds and attend to the magic of our radical embedding with other life-forms. Ultimately, dark ecology appears as an apophatic thought: letting the ontological uncanniness of our radical entanglement with other beings disclose itself, it remains open to an irreducible inaccessibility, a distance coinciding with a radical intimacy—an absence so near it infuses our whole being in the world. This apophatic cloud of unknowing grows with the consciousness of our earthly embeddedness: the more we learn about each other, the uncannier our intimacy becomes. Acknowledging that "nonhumans are installed at profound levels of the human—not just biologically and socially but in the very structure of thought and logic" thus demands an epistemological humility ready to embrace the ambiguities and uncertainties of our entangled condition. Morton refers to the philosophical love of wisdom as a "strange openness" (*DE*, 159, 92). With dark ecology, they summon ecological thought to nurture this strange openness and offer an *eco-concept* to celebrate the dark wonder at the heart of our being in the world.

NOTES

1. Jacques Derrida coined this neologism in *Spectres of Marx* to refer to communism's tendency to haunt Western societies even after the triumph of capitalism heralded the so-called end of history. Hauntology thus evokes the manifestation of a lingering ontology, the expression of an immaterial, persistent presence pervading narratives and imaginaries.
2. Thereafter *ET.*
3. Thereafter *EwN.*
4. Thereafter *DE.*
5. Morton still acknowledges the imperative to think about the concept of species differently, including the need to think humankind as a planetary totality without ceding to "hubristic technocracy" or to "the soppy and oppressive universalism and difference erasure that it usually implies" (*DE*, 22).
6. This also parallels the concept of Plantationocene, which identifies a hegemonic culture of extractive and enclosed plantations relying on diverse forms of exploitation and alienation, such as globalized factory meat production or monocropping farming.

REFERENCES

Abram, David and Strand, Sophie. "Magic as Radical Embedding in Our Web of Relations," October 31, 2022. https://www.youtube.com/watch?v=yC8m2-v4SwE.

Bahaffou, Myriam and Gorecki, Julie. Foreword. *Le Féminisme ou la mort*, by Françoise d'Eaubonne. Le Passage Clandestin, 2020.

Fressoz, Jean-Baptiste, *L'Apocalypse joyeuse. Une histoire du risque technologique.* Seuil, 2012.

Haraway, Donna. *Staying with the Trouble: Making Kin in the Chthulucene.* Duke University Press, 2016.

Irigaray, Luce. *This Sex Which Is Not One.* Cornell University Press, 1979.

Latour, Bruno. *Chroniques d'un amateur de sciences.* Presses des Mines, 2006.

———. *Face à Gaïa. Huit conférences sur le nouveau régime climatique.* La Découverte, 2015.

Malm, Andreas. *Fossil Capital. The Rise of Steam Power and the Roots of Global Warming.* Verso Books, 2016.

Morton, Timothy. *Dark Ecology: For a Logic of Future Coexistence.* Columbia University Press 2016.

———. *Ecology without Nature: Rethinking Environmental Aesthetics*. Cambridge: Harvard University Press, 2007.
———. “The Dark Ecology of Elegy,” in Karen Weisman (ed.), *The Oxford Handbook of the Elegy*, 2010.
———. *The Ecological Thought*. Harvard University Press, 2010.
———. “The Mesh” in Stephanie Lemenager (ed.), *Environmental Criticism for the Twenty-First Century*. Routledge, 2011.
———. “What Is Dark Ecology?” in *Living Earth: Field Notes from the Dark Ecology Project, 2014–2016*. Sonic Acts Press, 2016, pp. 29–56.
———. “Why Ecological Awareness Is Loopy,” in Solvejg Nitzke and Nicolas Pethes (ed.), *Imagining Earth: Concepts of Wholeness in Cultural Constructions of Our Home Planet*. Transcript Verlag, 2017, pp. 91–112.
Myers, Natasha. “How to Grow Livable Worlds: Ten Not-So-Easy Steps for Life in the Planthroposcene,” ABC Religion and Ethics. Australia Broadcasting Corporation, 2021.
Serres, Michel. *Le Contrat Naturel*. Flammarion, 2020.
Stengers, Isabelle. *In Catastrophic Times: Resisting the Coming Barbarism.* Open Humanities Press, 2015.

Chapter 4

Anthropocene and Capitalocene

Humans and Systems in Earth's Emergent Epoch

Rebekah A. Taylor-Wiseman

ANTHROPOCENE AS ECO-CONCEPT

The term, the *Anthropocene*, is a critical eco-concept because it gives name and coherence to a range of phenomena and measurements that are associated with global climate change and environmental collapse. Naming this emergent epoch after human, *Anthropos*, represents the awareness that humans are capable of impacting earth's systems on a scale similar to a super volcano or comet, as the emergent epoch is marked by exponentially increased CO_2 in the atmosphere, ocean acidification, extinction events, and loss of biodiversity. For some, the Anthropocene is an awakening and a reckoning of our failures and guilt, while others see in the name only a continuation of the hubris and binary thinking that fuels environmental degradation; of course, some say, humans would imagine themselves the center of everything in name and deed. Still, the most significant opposition to the term, Anthropocene, is that it perpetuates white, European, colonial historical narratives and signifies only the agency of a very small subset of humans, to the detriment of more complex thinking about (racialized) economic and historical systems. Not only is *Anthropos* synonymous with "white man" at its root, a "human age" falsely implicates all of humanity when, in fact, only a small percentage of the Earth's population drive climate change. In an interview with Gregg Mitman, Donna Haraway put it this way:

> There's a way in which the Anthropocene is considered a species act as opposed to an historical, situated set of conjunctures that are absolutely not a species act. Most peoples on this planet have precisely not lived and exercised the same kinds of processes that break generations, that radically simplify ecologies, that drastically force labor in a mass way that creates a kind of global transformation and global wealth that is in and of itself genocidal and extinctionist. That is not a species act; it's a situated historical set of conjunctures, and I think to this day the term Anthropocene makes it harder, not easier, for people to understand that. (Mitman n.p.)

As Haraway identifies global wealth as the culprit in the above passage, an alternative to the Anthropocene might more pointedly identify and condemn the economic system—namely, capitalism—that influences human behavior and drives environmental exploitation, a reasoning that led world-ecology theorist Jason Moore to coin the term Capitalocene. Our conceptualization and applications of this eco-concept, which names our emergent epoch becomes indicative of all we believe about humans and nature, then, as we attempt to explain, define, and categorize the human relationship with the earth and come to terms with the legacy that is written in the fossil layer.

ARRIVING AT THE ANTHROPOCENE

The Anthropocene replaces the Holocene—the unusually stable period of about the last 11,000 years that supported the development of sophisticated agricultural practices and allowed for massive, and recently rapid and destructive, human population growth. The story goes that Dutch chemist, Paul Crutzen, who shared a Nobel Prize for discovering the effects of ozone-depleting compounds coined "Anthropocene" while sitting in a meeting (Lawrence n.p.). Crutzen supposedly blurted out "We are no longer in the Holocene; we are in the Anthropocene" due to his observation that people have altered the composition of the atmosphere and Earth's geology (Kolbert 107–109). Crutzen then published "The Geology of Mankind" in *Nature* in 2002, and "Anthropocene" began appearing in scientific journals soon after, eventually entering popular usage. The Anthropocene Working Group (AWG) of the International Commission on Stratigraphy (ICS) voted in favor of Anthropocene as a geological time (chronostratigraphic) unit to indicate that we have moved out of the Holocene epoch into a new era based on measurable changes to the Earth's systems, but the AWG continues to debate over when the Anthropocene started and other key questions. The issue of the "golden spike"—or the moment when the new epoch is registered in the rock record typically lands in the twentieth century when phenomena that collectively indicate the Anthropocene such as the proliferation of new materials like concrete and plastics, erosion, freshwater usage, animal domestication, and deforestation increase rapidly. The Great Acceleration, as it is sometimes called, is often pinpointed to 1945–1950 when artificial radionuclides from nuclear bombs marked the fossil record in the "sharpest and most globally synchronous" of the Anthropocene signals (AWG).

For some, the Anthropocene was set in motion in 1784 with the invention of the steam engine, though Jason Moore cautions against the need for a more complex narrative of coal and steam as entangled with class and imperialism. The Capitalocene—signifying relations of power, nature, and capital—crystallized with Christopher Columbus, Moore insists (Patel and Moore 161–179). Some anthropologists push the conversation even earlier, noting the impacts of indigenous cultures and early agricultural practices that have left their mark on the land, though it is doubtful that pre-modern practices will be registered in the fossil layer long term. The invention of fire, some say, held the "spark" of combustion engine.

As denial, or social amnesia, goes hand in hand with climate change conversations, it's important to note that scientists were speculating and forewarning about human impact on the Earth long before the twentieth century, and much of the science that led to environmentally significant innovations was already in the works during the early twentieth century; radiometric dating was first proposed to measure the age of the Earth sometime around 1904, for example (Elias n.p.). Additionally, concerns over ozone depletion led Gordon Dobson of Oxford University to build the first instrument for measuring total ozone from the ground in the late 1920s (the Dobson unit or DU is now a measure of ozone concentration) (Somerville 127). The Keeling Curve, the definitive measure of carbon dioxide in the atmosphere, was put into commission in 1958 but provides a visual representation of the Anthropocene that extends beyond its official name. Using ice-core data, scientists are able to demonstrate the increase in atmospheric carbon dioxide using a persuasive line graph. Recently, the parts per million of atmospheric carbon dioxide surpassed the threshold of 400 ppm for the first time in recorded history (see https://keelingcurve.ucsd.edu/ for the latest measurements). These graphic depictions are crucial because, as we have noted, the Anthropocene is not otherwise directly perceivable (in its entirety, though we see the "symptoms"). Still, extremely detailed scientific reports are even now sometimes met with denial of anthropogenic impact, so it could be argued that cultural attitudes about nature and the human are just as, if not more, important as the quantitative measurements.

The etymology of *Anthropocene* takes us to the late nineteenth century, for the term is a variation of Italian geologist Antonio Stoppani's (1824–1891) use of "anthropozoic" to describe the time lapse of human induced change

(Zalasiewicz 835). Swedish Nobel Prize winning chemist Svante Arrhenius first theorized that the greenhouse effect could be enhanced by human activity in an 1896 publication, "On the Influence of Carbonic Acid in the Air upon the Temperature of the Ground" (Somerville 35). Arrhenius, along with Guy Stewart Callendar, was the first to suggest that burning increasing amounts of coal, oil, and fossil fuel would result in a measurable increase in atmospheric carbon dioxide (Anderson, Hawkins, and Jones 179), but his investigations were practically ignored and work completely forgotten by the 1970s. The carbon dioxide problem is treated as novel in 1970 publications, and again and again is treated as novel in each following decade, so that even now this information has an impression of discovery to the general public despite the fact that it is not so very new at all. Although the ideas "received short shrift in the geological community, seeming absurd when set aside the vastness (newly realized, also) of geological time," there were several precursors to the Anthropocene (Zalasiewicz, Williams, Steffen, and Crutzen 2228).

As another example of early twentieth century Anthropocenic awareness, consider the opening sentence of Nathaniel Southgate Shaler's 1905 book, *Man and Earth*: "The situation of man with reference to the material resources of the earth deserves more attention than has been given to it" (1). Shaler continues, "[w]e may be sure that those who look back upon us and our deeds from the centuries to come will remark upon the manner in which we use our heritage, and theirs, as we are now doing, in the spend-thrift's way, with no care for those to come" (1). Shaler's book is now out of print (but digitized in the free Internet Archive) while we think and write with the exact same sentiment over a century later.

George Perkins Marsh also anticipated current dialogues about the Anthropocene in his key text on the intersection of human time and geological time. Marsh's *Man and Nature* was first published in 1864 under the alternate title of *Physical Geography as Modified by Human Action*. Although the text fell out of popularity (because of Marsh's "savant-like approach," fluency in twenty languages, and subtitles such as "*Extirpation of Quadrupeds*") and only returned to print in 2003, it was the "most influential text of its time next to Darwin's *On the Origin of Species*" (Lowenthal xv). Even one hundred and fifty years ago, Marsh warned about the cumulative effect of human impact as an "interrelated whole": "Anyone who wields an ax knows its likely impact, but no one before George Perkins Marsh had gauged the cumulative effects of all axes—let alone chainsaws" (Lowenthal xv). Likewise, in the 1885 edition of *Man and Nature,* Marsh writes, "The immediate effects in such cases now elude human understanding, but who shall say that the mathematics of the future may not compute the measure of such agency and calculate even these small cosmical results of human action?" (Marsh 465, footnote). Marsh's insistence on a "cumulative" approach is picked up by Timothy Clark in this century, as he identifies the scale of human impact as a crucial point in conceiving the Anthropocene: "even trivial personal decisions about food, ways of travelling to work, gardening etc. all become significant or not depending on the contingency of how many others have done, are doing or will do them, anywhere on earth, implicating acts of seeming irrelevance in incalculable impacts" (Clark, "Climate Change Ironies" 136).

Lurking behind all of these conversations is the reality of competing timescales—the human and the geological. Humans are a "blip" when considered from a deep time perspective, and even the significant impact of anthropogenic (or capitalogenic) climate change on life on Earth (and all the suffering it causes) will hardly be registered in the Earth's fossil record when we're extinct. The "temporal conundrum" (Shoshitaishvili 135) of Anthropocene conversations parallels the competing conclusions that arise depending on whether the Anthropocene is thought of in geological or philosophical terms. Will the fossil layer tell a story of our grief, melancholy, and guilt? Will empire and colonialism be evident if there are anthropologists in the next eon?

THE CASE OF THE CAPITALOCENE

Jason Moore, author of Anthropocene *or Capitalocene?: Nature, History and the Crisis of Capitalism* (2016), has been the most insistent about designating the emergent epoch "Capitalocene" rather than "Anthropocene" to denote the "age of capital"-"the historical era shaped by the endless accumulation of capital" (Moore, "The Capitalocene, Part I" 3). The Capitalocene names the system that unevenly distributes agency and development, establishing nature as external and separate from human life. Moore's key concept of cheapened nature—easily accessible, consumable resources—points to the ways in which capitalism is self-exhausting and subject to nature's materiality.

For Moore, the crisis we are experiencing is "not a failure of a species, it's the failure of a system" and Anthropocene discourses blame the victims ("Capitalocene and Planetary Justice" 49). Instead of homogenous, abstracted "humanity," Moore argues, those responsible for the climate crisis "have names and addresses, starting with the eight richest men in the world with more wealth than the bottom 3.6 *billion* humans" (50, italics in original). In this blaming of the "richest men," though, Moore has already shifted his focus from the production and consumption systems of capitalism, back to the prototypical *Anthropos*, and while Moore is correct to insist that the majority of humans are victims not culprits, there must be some widespread accountability for fossil fuel consumption that powers electricity and an estimated 1.2 billion motor vehicles in use on Earth, for example.

In addition, while global population growth undeniably strains resources, the uneven, disproportionate impact of the "first world" cannot be overstated. Population growth rhetoric can falsely implicate those countries with the most negligible carbon footprint, when the reality is that the U.S., China, and Europe (regions with relatively low reproduction rates) have essentially caused global warming. Indeed, the carbon emissions of the richest 1 percent in the world is more than double the carbon emissions of the lowest 50 percent (Oxfam), and this fact is punctuated by the stark reality that the poorest on the earth, those least responsible for the so-called Anthropocene, suffer most from its manifestations.

Certainly, capitalism has fundamentally transformed nature and human nature as Marx predicted it would, but the complexity of global warming and all that the emergent phenomenon signals cannot be oversimplified into one easy culprit. Even as obvious a scapegoat as capitalism is, the legacy of capitalism is still being written, and things like green spaces and renewable energies are a product of capitalism too. To name an epoch while living in it is unusual for a practice that typically has the advantage of hindsight, and it is simply unknown what capitalism's legacy (in the fossil record) will be. Compounding the naming and dating issues that confound scientists (see Perkins, "Researchers move closer to defining the Anthropocene"), scholars in the Humanities like Timothy Morton cite the size and complexity of global warming in relation to human perception, which results from our situatedness *inside* of this "hyperobject"—a phenomenon "massively distributed in time and space relative to humans, and which defy overview and resist understanding" (Morton 1).

Even more, Indian novelist Amitav Ghosh argues that empire and imperialism are *equally* important drivers of global warming and must be considered alongside capitalism to gain a fuller (non-Eurocentric) picture of the Earth's transformation (87). Ghosh's analysis puts Asia, not the West, as central to the climate crisis, and while his argument is too rich to do justice to here, it is important to note the "paradoxical possibilities" of one central point that is related to the naming of our new epoch: the fact that some of the key technologies of the carbon economy were first adopted in England, the world's leading colonial power, may actually have *retarded* the onset of the climate crisis" (Ghosh 110, emphasis mine). In other words, capitalism and its twin empire, are responsible for the suppression of certain economies, which Ghosh argues, might have *delayed* the effects of global warming. Do frameworks such as this complicate the legacy of capitalism from a geological perspective? In Ghosh's study, capitalism is hardly the undisputed, unidirectional driver that "Capitalocene" proponents suggest, and perhaps perpetuates the "strikingly linear view of history" that Moore sets out to disrupt (14).

CONCLUSION

While academics debate the terminology, and in spite of its shortcomings, the Anthropocene has already entered popular usage. As an eco-concept, Anthropocene might seem insufficient, but it is productively flawed, as even critiques like Moore's invite conversations that identify injustices and encourage more responsible stewardship of Earth's resources. Since the "popular" Anthropocene is likely here to stay, we might follow Emma Dabiri's model in *What White People Can Do Next* and "thin[k] about the ways in which a vast array of oppressions or forms of disadvantage might have a common origin, in order to identify ways of coalition building that can focus on the source of the problem, while remaining mindful of the different textures of our varied but interconnected struggles" (41). Dabiri's model of combatting white supremacy is instructive here too, and actions such as 1) stop the denial, 2) interrogate whiteness, 3) abandon guilt, 4) redistribute resources, and 5) "recognize this shit is killing you too" are a productive place to start (Dabiri 130). How, though, to convince people that capitalism *does not* work for them?

Moore is critical of the Popular Anthropocene, but climate change solutions will require the cooperation of capitalists and the general public, and the populist trends in global politics over the last decade have demonstrated that capitalism is still associated with freedoms and opportunity in the zeitgeist of the leading economic powers. The "naming debate" potentially delays action when there is no time to waste. A critical Anthropocene is possible that includes capitalism as an ecology (Moore) and works toward action on behalf of those most at risk of suffering due to global climate change. Rob Nixon, author of *Slow Violence and the Environmentalism of the Poor*, summarizes the task at hand this way in his piece "The Great Acceleration and the Great Divergence: Vulnerability in the Anthropocene":

> So a crucial challenge facing us is this: how do we tell two large stories that can often seem in tension with each other, a convergent story and a divergent one? First, a collective story about humanity's impacts that will be legible in the earth's geophysical systems for millennia to come. Second, a much more fractured narrative. For the species-centered Anthropocene meme has arisen in the twenty-first century, a period in which most human societies have experienced a deepening schism between the uber-rich and the ultra-poor. In terms of the history of ideas, what does it mean that the Anthropocene as a grand explanatory species story has taken hold during a plutocratic age? And from an imaginative perspective, how can we counter the centripetal force of the dominant Anthropocene species story with centrifugal stories that acknowledge immense inequalities in planet-altering powers? (n.p.)

The Anthropocene is a lens and frame through which to consider these questions and more. In its controversy is its value—to inspire conversation and debate—but we cannot be distracted from the real threat, destruction of our biosphere, which will (eventually) affect all life on earth regardless of status. Arguably, the Anthropocene concept has already increased environmental awareness by indicating the magnitude and speed of (some) humans' impact, as well as indicating a tipping point where that impact will be irreversible. The outpouring of interdisciplinary scholarship on Anthropocene (*Anthropocene Feminisms*, *Mourning in the Anthropocene*, *The Anthropocene Unconscious*) gives it robust pedagogical and practical applications. Perhaps most importantly, discussions of Earth's time have the potential to humble and resituate us to consider subjectivities and realities that extend beyond our own bodies, which might just inspire care and concern for others who coexist with us now and those to come.

REFERENCES

Anderson, Thomas R., Ed Hawkins, and Philip D. Jones. "CO2, the Greenhouse Effect and Global Warming: From the Pioneering Work of Arrhenius and Callendar to Today's Earth System Models." *Endeavour,* vol. 40, no. 3, 2016, pp. 178–187, https://doi.org/10.1016/j.endeavour.2016.07.002

Anthropocene Working Group (AWG). "Results of Binding Vote by AWG." Subcomission on *Quaternary Stratigraphy*. May 21, 2019. http://quaternary.stratigraphy.org/working-groups/anthropocene/

Clark, Timothy. *Ecocriticism on the Edge: The Anthropocene as a Threshold Concept.* Bloomsbury, 2015.

Elias, Scott A. "History of Quaternary Science" *Reference Module in Earth Systems and Environmental Sciences*. Elsevier, 2015. https://doi.org/10.1016/B978-0-12-409548-9.09548-8

Ghosh, Amitav. *The Great Derangement: Climate Change and the Unthinkable*. University of Chicago Press, 2016.

Kolbert, Elizabeth. *The Sixth Extinction.* Henry Holt, 2014.

Lawrence, Mark. "Remembering Paul Crutzen: A Grand Scientist who took on the Grand Challenges of the Anthropocene." *C2G*. February 23, 2021. https://www.c2g2.net/remembering-paul-crutzen-a-grand-scientist-who-took-on-the-grand-challenges-of-the-anthropocene/

Mitman, Gregg. "Reflections on the Plantationocene: A Conversation with Donna Haraway and Anna Tsing," *Edge Effects*, https://edgeeffects.net/haraway-tsing-plantationocene/

Moore, Jason. *Anthropocene or Capitalocene?* PM Press, 2016.

———. "The Capitalocene, Part I: On the Nature and Origins of Our Ecological Crisis." *The Journal of Peasant Studies*, vol. 44, no. 3, 2017, 594–630. doi: 10.1080/03066150.2016.1235036.

———. "Capitalocene and Planetary Justice." *Maize*, vol. 6, 2019, pp. 49–54.
Morton, Timothy. *Hyperobjects: Philosophy and Ecology after the End of the World.* University of Minnesota Press, 2013.
Nixon, Rob. "The Anthropocene: The Promise and Pitfalls of an Epochal Idea." https://edgeeffects.net/anthropocene-promise-and-pitfalls/
———. "The Great Acceleration and the Great Divergence: Vulnerability in the Anthropocene." *MLA Profession*, 2014. https://profession.mla.org/the-great-acceleration-and-the-great-divergence-vulnerability-in-the-anthropocene/
Oxfam. "Carbon Emissions of Richest 1 Percent." September 21, 2020. https://www.oxfam.org/en/press-releases/carbon-emissions-richest-1-percent-more-double-emissions-poorest-halfhumanity#:~:text=The%20richest%20one%20percent%20of,period%20of%20unprecedented%20emissions%20growth
Patel, Raj and Moore, Jason W. *A History of the World in Seven Cheap Things: A Guide to Capitalism, Nature, and the Future of the Planet.* University of California Press, 2017.
Perkins, Sid. "Researchers Move Closer to Defining the Anthropocene." *Proceedings of the National Academy of Sciences–PNAS*, vol. 120, no. 29, 2023, pp. e2310613120–e2310613120. https://doi.org/10.1073/PNAS.2310613120
Shoshitaishvili, Boris. "Deep Time and Compressed Time in the Anthropocene: The New Timescape and the Value of Cosmic Storytelling." *The Anthropocene Review*, vol. 7, no. 2, 2020, pp. 125–137. https://doi.org/10.1177/2053019620917707
Somerville, Richard C.J. *The Forgiving Air.* University of California Press, 1996.
Zalasiewicz, Jan, Mark Williams, Alan Haywood, and Michael Ellis. "The Anthropocene: A New Epoch of Geological Time?" *Philosophical Transactions of the Royal Society*, vol. 369, 2011, pp. 835–841. doi: 10.1098/rsta.2010.0339.
Zalasiewicz, Jan, Mark Williams, Will Steffen, and Paul Crutzen. "The New World of the Anthropocene." *Environmental Science & Technology*, vol. 44, no. 7, 2020, pp. 2228–2231. doi: 10.1021/es903118j.

Chapter 5

Plantationocene, Chthulucene, and Agency as Process and Relation

Lenka Filipova

In recent years, two eco-concepts which provide a lens for understanding the intertwined environmental and social challenges of contemporary time have gained traction in the environmental humanities: the Plantationocene, a term which emerged from a collective discussion published in the journal *Ethnos* (Haraway et al. 2016), and the Chthulucene, coined by Donna Haraway (Haraway 2016a). The Plantationocene links extractive practices, monoculture, and forced labor to modernity and climate change since the 1600s. It emphasizes imperialism's lasting impact on social hierarchies and inequalities based on gender and race. The Chthulucene, which may be understood as a complementary notion to Plantationocene, serves as a framework for contemplating the interconnectedness of human and more-than-human entities amidst the devastating loss of life and environmental degradation. These environmental concepts provide valuable tools for grappling with the intertwined environmental and social challenges of our time as they examine the ways in which human and more-than-human actors interact in ecological change and consider the uneven ways in which these changes impact different populations and regions.

The emergence of both notions represents a profound critique of the limitations of the Anthropocene and Capitalocene frameworks, highlighting the need for a more nuanced and inclusive understanding of the entangled relationships between humans and more-than-humans. The Anthropocene and Capitalocene have been criticized for promoting cynicism and defeatism due to their focus on the negative impacts of human activity on the planet. While the Plantationocene foregrounds the imperialist factors, which have resulted in unequal eco-social degradation, Haraway offers the Chthulucene as a conceptual framework that emphasizes ongoing multispecies stories and practices of becoming-with in times of uncertainty and precarity. As Haraway puts it, "the biotic and abiotic powers of this earth are the main story" in the Chthulucene, where humans coexist with other beings as integral parts of the earth (Haraway, 2016b). In order to illuminate the emergence and significance of both the Plantationocene and Chthulucene, it is essential to provide a succinct overview of the critique of the Anthropocene.

Since the term Anthropocene was popularized by the atmospheric chemist Paul Crutzen in the early 2000s, it has garnered considerable popularity among scientists as well as criticism. Critics of the concept have raised a central question regarding the collective subject that is said to have become a geophysical force on a planetary scale, as it has remained unclear who or what that collective subject is (Crist 2003; Chakrabarty 2009). The category of the Anthropos and the human is not simply a descriptive term or a timeless reality, but rather a construct that has historically conferred certain privileges on those who identify with it, while simultaneously constituting exclusions. Scholars have therefore criticized the term by highlighting the historical contingency of the human. In other words, notions of the human have always relied upon notions of the nonhuman and inhuman. Even the concept of the posthuman, offered most recently as an alternative to the human, therefore needs to be seen as part of particular constellation of eco-social and political relations, and as such, is not epistemologically innocent or necessarily more inclusive.

Recently, feminist and postcolonial critics have argued for the need to deconstruct the imagined homogeneity of the Anthropos by highlighting how the so-called Age of Humanity is enacted differently across temporalities, spatialities and bodies. These contestations have generated a number of reformative reconfigurations of the concept (see, for example, Bennett 2010; Parikka 2015; Alaimo 2016; Colebrook 2016; Morton 2016). Both the Plantationocene and the Chtuhulucene are two especially influential recent configurations of this critique. Scholars have increasingly argued that the plantation and its legacy are central to understanding the modern era, and have embraced the concept of the Plantationocene (Aikens et al. 2019; Davis et al. 2019; Carney 2020). The concept builds on previous research on the unique characteristics of *plantation economies* (Beckford 1972; Mintz 1986; Woods 1998), and posits that large-scale, export-oriented agriculture reliant on the simplification of landscapes, the estrangement of land and labor, the movement of genomes, plants, animals, and people, as well as coerced labor has been a fundamental force in shaping modern life since the arrival of European power in the Americas, Asia, and Africa. The legacy of the plantation system is thus deeply entwined with issues of race and power, and the Plantationocene can be understood as a name for the geological epoch that is shaped by the history of colonialism and slavery, and that in turn shapes the present and future of racial capitalism. This history has left a profound mark on the landscape, as well as on the bodies and psyches of those who have been oppressed by it, and the notion of Plantationocene helps to make visible power relations and social inequalities that define "the Anthropocene as a racial process," in both historical and spatial terms (Pulido 2018).

In the following I will look at some of the iterations of both Plantationocene and Chthulucene. Afterwards I will reflect on the awareness of the inescapable connectedness of planetary existence, and the potential political significance of this awareness. I will argue that more attention needs to be drawn to the agency of the more-than-human if we want to overcome the separation of human and more-than-human political actors.

PLANTATIONOCENE

According to Haraway the Plantationocene marks a significant turning point in human-mediated history with the emergence of plantation economies in the sixteenth century. Haraway highlights the importance of considering the sympoetic elements within plantation processes. She suggests that it is essential to discuss the intricate networks connecting sugar, precious metals, plantations, Indigenous genocide, and slavery, which involved labor innovations and relocations that affected both human and more-than-human workers. This recomposition of "critters and things" underscores the importance of understanding the complex entanglements of all involved (Haraway, "Tentacular" 48). In addition, Anna Tsing posits that plantation logics are distinguished by their "scalability" (38) and "interchangeability" (39). Scalability refers to the plantation's ability to expand rapidly by following a predetermined blueprint that involved the displacement of local people and plants, clearing of land, and importation of foreign crops and labor. Interchangeability stands for the ease of replacing one species with another, exemplified by the plantation's practice of substituting cane stock for enslaved individuals (Tsing 38–39). By highlighting the crucial role of plantation ecologies and politics in shaping the present, both Tsing and Haraway offer a way to challenge the Eurocentric narrative that places coal, the steam engine, and the industrial revolution at the epicenter of global environmental change. By decentering the Eurocentric perspective, they allow for a consideration of alternative viewpoints regarding the causes and consequences of environmental change.

In order to provide a more nuanced understanding of the relationships between land, labor, and power in the history of colonialism, slavery and global environmental change, Katherine McKittrick has emphasized the significance of the plantation as a "very meaningful geographic prototype that not only housed and normalized (vis-à-vis enforced placelessness) racial violence in the Americas but also naturalized a plantation logic that anticipated (but did not twin) the empirical decay and death of a very complex black sense of place" (951). By recognizing the centrality of the plantation in shaping modernity and coloniality, McKittrick demonstrates how the concept of the Plantationocene can help to center discussions of race and the ongoing impacts of colonialism on marginalized communities and a black sense of place (951).

The ongoing discussion surrounding the Plantationocene is still evolving and thus far has been characterized by a certain tentativeness, with scholars offering varied perspectives and critiques. Among them, Davis et al. (2019) have compellingly argued that the environmental humanities approach, while shedding light on ecological disruptions, has largely sidelined issues of racism. The authors emphasize that the violence of enslavement and the construction of a new, race-based world order have been central to the formation of the Plantationocene, and must therefore be accorded greater attention in scholarly discourse. They contend that such an approach is essential to fully comprehend the ways in which racism, capitalism, and environmental degradation intersect and impact marginalized communities. Relatedly, Aikens et al. (2019) have emphasized the intersectionality of race and colonialism. Drawing on critical literary theory and the notion of the Black Atlantic (Gilroy 1993), the authors highlight the ways in which "plantation ideologies, iconographies and narratives continue to structure everyday life" (Aikens et al.). By foregrounding the role of plantation ideologies, the authors highlight the ways in which the legacies of colonialism are not only economic and ecological but also cultural and psychological. They also call for greater attention to the lived experiences of those who have been most directly affected by plantation economies and their legacies. This includes not only enslaved Africans but also Indigenous peoples, women, and other marginalized groups whose labor and bodies were commodified and exploited under colonialism. A more inclusive approach to the Plantationocene, according to them, must center the perspectives and experiences of marginalized groups to understand the continued influence of plantation economies on social and environmental relations and engage with questions of power, identity, and justice.

More recently, Judith A. Carney has further developed the idea of the Plantationocene by incorporating Sylvia Wynter's 1971 work on the link between the rise of the novel and plantation economy. Carney draws on Wynter's critical analysis of the plantation economy to illustrate the profound devastation caused by the plantation-based slave trade. At the same time, however, she highlights the possibilities for resistance and resilience in the cultivation of small garden plots and smallholder farming systems for both cultural and physical sustenance. Her approach sheds light on the often-overlooked agency and resistance of enslaved peoples who were able to maintain their cultural and spiritual traditions despite the brutal oppression they faced. As Carney argues, this form of independent agricultural production provided "a strategy for expanding social networks beyond the plantation" and "was a crucial step towards asserting new rights, organizing resistance, and gaining leverage in an exploitative system" (1083–84). By focusing on the ways in which enslaved peoples adapted to and resisted the plantation system, Carney challenges the notion that they were passive victims of historical forces beyond their control and makes an important contribution concerning identity and agency in and beyond the Plantationocene.

CHTHULUCENE

While the Plantationocene highlights the legacies of colonialism and the Capitalocene's neglect to take responsibility for its participation in oppressive systems such as slavery, genocide of Indigenous communities, and the forced displacement of humans and more-than-humans, Chthulucene speaks to the distributed agencies involved in extinction and climate change. Haraway uses this term in *Staying with the Trouble: Making Kin in the Chthulucene* to emphasize the importance of thinking beyond the human-centric perspectives "past, present, and to come" (101), and in order to recognize the interconnectedness of all beings and entities. At the same time, she endeavors to find a way between catastrophism on the one hand, and a naïve faith in technological fixes on the other, both of which, as she argues, characterize the Anthropocene. She argues that the Anthropocene implies that it is too late to change the future and that such an approach demands a deliberate disregard, a conscious negation, and a refusal to accept reality. She presents an alternative in the form of a response that involves acknowledging the challenges and embracing them wholeheartedly, fostering connections with other species to tackle these issues collectively, and capturing the dynamic "ongoingness" of powers and forces of which people are only a part (3). The term itself comes from the Greek *khthonios*, which she translates as "'of, in, or under the earth and the seas'—a rich terran muddle for SF [speculative fabulation], science fact, science fiction, speculative feminism, and speculative fabulation" (53). Yet it is also partially derived from "tentacular arachnid" (31), the Pimoa *Cthulhu*, another-than-human guide to tentacular thinking, which stands for thinking nets and networks, and life lived along lines and interlaced trails.

More specifically, Haraway explores the concept of reciprocal relationality, which could potentially result from the collective activism of multiple species, and the regenerative potential of multispecies or "naturalcultural assemblages" (*Chthulucene* 38). She weaves a narrative that acknowledges the crucial role of Indigenous communities in creating positive change, but also integrates various fields of study such as evolution, ecology, history, and cosmology. By recognizing the agency of more-than-human actors and highlighting the need for multispecies collaboration to address environmental challenges, Haraway emphasizes that achieving resurgence requires imagination and advocating for transformative learning to become more "response-able" (69), that is, capable of living with other species on an ecologically compromised planet.

Haraway's approach significantly draws on Indigenous knowledges, such as those of the Hopi, the Navajo, or those in Madagascar and the North American circumpolar lands, which, although regionally different, tend to foreground infinite relationality of human and more-than-human actants. To take an example from a different continent, Deborah Bird Rose explains that in Indigenous Australia, the term "Country" (7), that is, the natural world and the land, is understood as both natural and social, and as a dynamic life process: "People say that country knows, hears, smells, takes notice, takes care, is sorry or happy . . . country is a living entity with a yesterday, today and tomorrow, with a consciousness, and a will toward life" (7). Similarly, Viveiros de Castro portrays the Amerindian worldview as a complex universe comprising numerous perspectives (55). In this framework, each more-than-human and human entity represents a focal point of awareness that perceives other entities based on their unique traits and abilities.

The eco-concepts of Plantationocene and Chthulucene reflect a growing consensus that the roots of global climate change can be traced back to the exploitation of the natural environment, as well as the subjugation of subaltern populations. These factors have been present since the inception of Western colonialism, and were further exacerbated by the industrial revolution in Europe. It follows that any efforts towards an emancipatory and reparative environmental politics must consider both human and more-than-human agency. However, contemporary environmental reforms have primarily focused on human relations of production, disregarding the political agency of more-than-humans. In the following discussion, I aim to emphasize the importance of recognizing this agency, as well as the challenges that arise in realizing the potential of tentacular thinking and fostering a more inclusive politics of human and more-than-human coexistence.

AGENCY AS PROCESS AND RELATION

Besides Haraway's Chthulucene, the agency of the more-than-human has been foregrounded by materialist scholars who do not only speak to the necessity to decolonize the Anthropocene, but also attend to the idea of power as "biogeotechno-power" (105) as proposed by Kathryn Yusoff, and to the element of caprice of the more-than-human in Fréderic Neyrat's notion of the "unconstructable Earth" (165).

According to Clark and Yusoff, to decolonize the Anthropocene does not only require contesting colonial geology and advocate for multiple origin stories structured along the relational spacetime of kinship rather than the linear time of the Anthropocene's ticking clock. It also needs to involve a different understanding of the relationship between geological forces and social practices, one that recognizes how agency is both constrained and made possible by "the forces of the earth" itself (Clark and Yusoff 5). In other words, Clark and Yusoff argue that the Anthropocene also needs to be understood as a geosocial formation. They suggest that it can be seen as a product of the relationship between social and geological forces, in which human activity and geological processes are mutually reinforcing. What is at issue is not only how to expand the composition of social words by reimagining how matter figures in them, but also how to respond to forces that have the potential to disrupt, weaken, or overpower the fundamental principles of governance and societal interactions. Clark and Yusoff argue that it is imperative to consider both aspects in order to establish a robust and sustainable framework for political and social systems. In this sense, they also suggest that geosocial formations, or the social and material relationships between humans and the environment, are crucial to understanding the Anthropocene.

In a similar way, Fréderic Neyrat suggests that caprice is a fundamental aspect of the Earth's more-than-human agency and that it poses a challenge to traditional notions of human power and control, including the use of technology. Neyrat is critical of geoconstructivist projects which are legitimized by the Anthropocene's rendering of the earth as an object and reliance on technofixes. He argues that this narrative is lacking precisely what it claims to have: "proper knowledge of the earth itself" (Neyrat 42). While humans have become geological agents, Neyrat argues that the earth has not, however, dissolved into humanity. He suggests that the "unnaturalist position" (65) underlying geo-constructivist politics negates the alterity of the more-than-human, creating an idea of humanity as abstract and off-planet. Therefore, stewardship should conserve some "part of alterity" within itself, thereby recognizing the difference between the human and more-than-human (58). This approach is important because, similar to Clark and Yusoff, it does not simply see the more-than-human as agential, but also foregrounds its unpredictability and power. As such, both Neyrat as well as Clark and Yusoff evince a salutary caution in squaring the more-than-human agency with human political agency, while also underpinning the difficulty of overcoming this separation while thinking about the interconnectedness of all life.

Yet I would like to suggest that the question of preserving alterity ceases to be as urgent as it may seem once more-than-human agency is not understood as an attribution (separate entities become "endowed" with agency), but as always already processual and relational. In the examples of Indigenous epistemologies briefly outlined above, as well as in many others, human and more-than-human entities are not given agency because they are *always already agential.* Tim Ingold explains that in Indigenous thought, the world is not simply invested with spirits that would give it life. People do not universally discriminate between living and non-living things: "for many people, life is not an attribute of things at all. That is to say, it does not emanate from a world that already exists, populated by objects-as-such, but is rather imminent in the very process of that world's continual generation or coming-into-being" (Ingold 67). Agency, as understood here, is not limited to human actions, but rather the result of a complex network of relationships between human and more-than-human entities. These actants constantly shape and create each other, making agency a dynamic potential that is inherently relational. The world, as a coalition of human and more-than-human life, is always in a state of flux, and nothing can act in isolation. Therefore, agency is always already relational because nothing can act on its own, and, as long as the world continues to exist, it is processual. As such, all we need to do is to be attentive to the presence of multiple agencies through which we move and which move us, which are dependent on us and which, much more importantly, we depend on. In such an approach, which may be more complex than it sounds, the idea that human agency is somehow supreme to other more-than-human agencies becomes obsolete.

While the concept of inherent agency in both human and more-than-human entities may appear straightforward, its implications are significant. This understanding demands that we recognize an agency as an essential aspect of their existence, underscoring the need for greater acknowledgement and respect of their agency. As such, it requires a shift in socio-political perspectives. Bruno Latour offers one such perspective by suggesting that the social needs to be understood "not as a special domain, a specific realm, or a particular sort of thing, but only as a very peculiar movement of re-association and reassembling" by which human politics is joined to more-than-human actants (7). Some political practices are currently underway that acknowledge the dynamic relationship between the human and more-than-human. For instance, recognizing the legal rights of nature is one such practice (see, for example, Solón 2018; Kauffman and Sheehan 2019). However, despite the positive intentions of these efforts, their full consequences remain uncertain and are yet to be fully assessed. It is worth noting that conferring rights onto the world can still reinforce the illusion of attribution, which limits our ability to fully comprehend the complex dynamics between human and more-than-human entities.

The eco-concepts of Plantationocene and Chthulucene are crucial for comprehending the environmental impact of colonialism. These concepts highlight historical and contemporary inequities and offer a perspective that views the world as a complex interplay between human and more-than-human actants. However, further scholarship is needed to explore the nuanced political implications and potential outcomes of recognizing more-than-human agency as a relational and processual force in achieving a sustainable future.

REFERENCES

Aikens, Natalie, Amy Clukey, Amy King, and Isadora Wagner. "South to the Plantationocene." *ASAP Journal*, October 17, 2019. http://asapjournal.com/south-to-the-plantationocene-natalie-aikens-amy-clukey-amy-k-king-and-isadora-wagner/.

Alaimo, Stacy. *Exposed: Environmental Politics and Pleasures in Posthuman Times*. University of Minnesota Press, 2016.

Beckford, George L. *Persistent Poverty: Underdevelopment in Plantation Economies of the Third World*. The University of the West Indies Press, 1972.

Bennett, Jane. *Vibrant Matter. A Political Ecology of Things*. Duke University Press, 2010.

Carney, Judith. "Subsistence in the Plantationocene: Dooryard Gardens, Agrobiodiversity, and the Subaltern Economies of Slavery." *The Journal of Peasant Studies*, vol. 48, no. 5, 2020, pp. 1075–99.

Chakrabarty, Dipesh. "The Climate of History. Four Theses." *Critical Inquiry*, vol. 35, no. 2, 2009, pp. 197–222.

Clark, Nigel, and Kathryn Yusoff. "Geosocial Formations and the Anthropocene." *Theory, Culture & Society*, vol. 34, no. 2–3, 2017, pp. 3–23.

Colebrook, Claire. "A Grandiose Time of Coexistence: Stratigraphy of the Anthropocene." *Deleuze Studies*, vol. 10, no. 4, 2016, pp. 440–54.

Crist, Eileen. "On the Poverty of Our Nomenclature." *Environmental Humanities* vol. 3, 2013, pp. 129–47.

Crutzen, Paul. "Geology of Mankind." *Nature*, vol. 415, 2002, p. 23.

Davis, Janae, Alex A. Moulton, Levi van Sant, and Brian Williams. "Anthropocene, Capitalocene, . . . Plantationocene? A Manifesto for Ecological Justice in an Age of Global Crises." *Geography Compass*, vol. 13, no. 5, 2019, pp. 1–15.

De Castro, Eduardo Viveiros. *Cannibal Metaphysics: For a Post-Structural Anthropology*, translated by Peter Skafish. Univocal, 2014.

Gilroy, Paul. *The Black Atlantic: Modernity and Double-Consciousness*. Verso, 1993.

Haraway, Donna. "Anthropocene, Capitalocene, Plantationocene, Chthulucene: Making Kin." *Environmental Humanities*, vol. 6, 2015, pp. 159–165.

———. *Staying with the Trouble: Making Kin in the Chthulucene*. Duke University Press, 2016.

———. "Tentacular Thinking: Anthropocene, Capitalocene, Chthulucene." *e-flux Journal*, no. 75, September 2016. https://www.e-flux.com/journal/75/67125/tentacular-thinking-anthropocene-capitalocene-chthulucene/.

Haraway, Donna et al. "Anthropologists Are Talking—About the Anthropocene." *Ethnos*, vol. 81, no. 3, 2016, pp. 535–64.

Ingold, Tim. *Being Alive: Essays on Movement, Knowledge and Description*. Routledge, 2011.

Kauffman, Craig and Linda Sheehan. "The Rights of Nature: Guiding Our Responsibilities through Standards." *Environmental Rights: The Development of Standards*, edited by Stephen J. Turner et al. Cambridge University Press, pp. 342–66.

Latour, Bruno. *Reassembling the Social: An Introduction to Actor-Network-Theory*. Oxford University Press, 2016.

McKittrick, Katherine. "On Plantations, Prisons and a Black Sense of Place." *Social & Cultural Geography*, vol. 12, no. 8, 2011, pp. 947–63.

Mintz, Sidney W. *Sweetness and Power: The Place of Sugar in Modern History*. Penguin, 1986.

Morton, Timothy. *Dark Ecology. For a Logic of Future Coexistence*. Columbia University Press, 2016.

Neyrat, Frédéric. *The Unconstructable Earth: an Ecology of Separation*. Fordham University Press, 2019.

Parikka, Jussi. *The Anthropocene*. University of Minnesota Press, 2015.

Pulido, Laura. "Racism and the Anthropocene." *Future Remains: A Cabinet of Curiosities for the Anthropocene*, edited by. Gregg Mitman, Marco Armiero, and Robert S. Emmett. University of Chicago Press, 2018, pp. 116–28.

Rose, Deborah Bird. *Nourishing Terrains: Australian Aboriginal Views of Landscape and Wilderness*. Australian Heritage Commission, 1996.

Solón, Pablo. "The Rights of Mother Earth." *The Climate Crisis: South African and Global Democratic Eco-Socialist Alternatives*, edited by Vishwas Satgar, Wits University Press, 2018, pp. 107–30.

Tsing, Anna L. *The Mushroom at the End of the World: On the Possibility of Life in Capitalist Ruins*. University of Princeton Press, 2015.

Yusoff, Kathryn. "Geosocial Strata." *Theory, Culture & Society*, vol. 34. no. 2–3, 2017, pp. 105–27.

Woods, Clyde. *Development Arrested: The Blues and Plantation Power in the Mississippi Delta*. Verso, 1998.

Chapter 6

Bioregionalism and Biocultural Region

Reconceptualizing the Human-Environment-Place Interrelationships Beyond the Culture/Nature Dichotomy

Abhra Paul & Amarjeet Nayak

Bioregionalism encourages intimate knowledge born out of dwelling in place, community, and local culture. The bioregional approach regards the region as a physical and cultural ecology of a place where ecological and cultural systems interact to shape one another (Bunting 57). The impact of humans in a specific bioregion and the study of the environment, species life, and climate condition are important factors to form the foundation of the sustainable culture that is the ultimate goal of bioregionalism. In the broadest sense of the eco-concept, bioregionalism focuses on the ecocentric side of environmental theories or concepts. As an eco-concept, it tends to challenge the human-centered frameworks of other ecocritical concepts harbouring hegemonic constructions about the place, cultural systems and natural systems (for example, dualism, domination of nature, Cartesian view/mechanistic view of the world, anthropocentric views, representation/construction of human/nonhuman agency, religious tradition and roots of present ecological crisis, etc. leading to an exaggerated sense of culture's role and underestimation of nature). Therefore, the *bioregional eco-concept* aims to generate a dialogue (the dialectical relationship between nature and culture) limiting the hegemonic splitting conceptual frameworks.

Scholars of ecocriticism perceive and construct a bioregional place as the "ecological context for transforming cultural relationships to nature"

(Ryan 80) and attempt to reimagine human connections to places. This chapter proposes to explore the theoretical perspective of bioregionalism, positing "place as a complex of nature and culture" (Ryan 81). It reconceptualizes human-environment interrelationships beyond the culture/nature dichotomy and considers that a place exists as a biocultural region. The interaction between the bioregion and human culture can directly influence and dismantle the privileges of any side over the other (culture/nature). The chapter presents the ecocritical scholarship on bioregionalism and culture/nature dualism. In this part, we outline major interpretations of Western thoughts on the nature-culture dichotomy and ecofeminist and ecocritical scholarship. We explain that these *seemingly separated* domains are challenged in the works of ecocritical theorists and literature related to bioregionalism and place studies. Our engagement here is, broadly speaking, to map elements of human culture intertwined with the dynamism of a particular bioregion: *a possible mix to re-read environmental literature*. The following part of the chapter will consider the theoretical perspective of the biocultural region, the *possible mix*, within the umbrella term of bioregionalism.

ECOCRITICISM AND ECOFEMINISM: THE NEGOTIATIONS BETWEEN NATURE AND CULTURE

In Western intellectual tradition, the culture/nature dualism is prominent and explained by many authors and ecofeminist scholars. This dualistic thought is identified with the mind/body split in the Cartesian worldview:

> The fact that different philosophers and different periods of philosophy have focused on different pairs of these dualisms and have defended different linking postulates has obscured the pervasiveness of dualistic and rationalist influence in philosophy. Thus Hegel and Rousseau emphasize the postulates linking public/private, male/female, universal/particular and reason/nature (Lloyd 58–63 and 80–85). For Plato the emphasis is mainly on reason/body, reason/emotion, universal/particular; for Descartes it is on mind/body (physicality), subject/object, human/nature and human/animal; for Marx it is on freedom/necessity, culture (history)/nature, civilized/primitive, mental/manual (a variant on mind/body) and production/reproduction. (Plumwood 45)

In modern Western thought, culture is radically situated in a hierarchical relationship of dominance to nature. In their works, ecofeminist scholars namely Val Plumwood, Karen Warren, and Carolyn Merchant have critiqued the structure of Western patriarchal culture and its connection to the ongoing domination of nature, women, nonhumans, and radically excluded humans, etc. Ecofeminism explains the framework that authorizes oppression (Gaard 2). It refers to ideologies being the root of dominance based on differences, such as the "logic of domination" (Warren 24), value and hierarchical dualisms and binary opposition (Plumwood 36–47) between male/female, culture/nature, superior/inferior, (as explained by Warren in *value hierarchical* (Up-Down thinking) and *value dualisms* (either-or thinking). The dualisms exemplify oppositional terms with a higher value attributed to the left term and inferior to the right term (Plumwood 47). Plumwood also states that nature has been systematically subordinated to the master subject that involves privileges of male over female and reason/culture over nature (42). Therefore, the subject-object relationship strengthens dualism.

Merchant explores the basic dichotomy between subject and object. She states that the "objectification of nature is rooted in Aristotle's locus of reality in the objects of the natural world" and made explicit in Descartes' "separation of mind from matter, that is, of thinking subject from external object" (60). Merchant further argues that culture is identified with the active subject and thus with the male. This basic philosophical framework of Western thought entails a philosophy of domination: an active controlling subject is dominant over and is separate from a passive controlled object and, thus, maintains hierarchical dominance of subject over object, male over female, and culture over nature (61).

Ecofeminist thoughts deal most comprehensively with dualistic thinking, control, and domination of nature at all levels. Ecofeminists point out that we live in a dominating culture that has compartmentalized our connection with nature. The image of the earth as inert, passive, and mechanistic leads to control and power over nature. The place is at the core of our contemporary severance from nature, and the concept of domination coalesces around the term bioregion. Digging the earth for minerals, clearing the forest for human settlements, and damming the river are various forms of oppression of nature. Such actions deviate from the key idea of bioregionalism: live-in-place sustainably and learn to become native to a place.

Doug Aberley states that bioregionalism pays consistent attention to the practical relationships with other social and environmental movements and thus welcomes intersections with values from ecofeminism, earth spirituality, permaculture, and ecological restoration. He quotes Michelle Summer Fike and Sarah Kerr: "bioregionalism and ecofeminism are two streams of the contemporary environmental movement that provide related yet distinct frameworks for analyzing environmental and social justice issues, as well as offering visions of more sustainable ways of living with the earth" (34). Judith Plant explains the common ground ecofeminism and bioregionalism share beyond the thinking in pairs of opposition (men/women, culture/nature): the human cultural requirements and needs to be sustainably adaptive within our nonhuman environments. Plant states that ecofeminism and bioregionalism attempt to rethink and rebuild human (culture) and natural communities as these two theories advocate decentralization of power and dismantle thinking in oppositional pairs. Bioregionalism, with its emphasis on distinct regional cultures attached to the natural environment, can provide a perfect framework to realize the full potential of ecofeminist philosophy (673–675).

The existing scholarly dialogue on dualism proves that it is not easy to negate it, but it is possible to create an alternate view beyond the dualistic framework. The recent ecocritical theories attempt to reconcile culture and nature without rejecting the differences and limits to foster ecological consciousness (an ecologically informed dialogue). This is what Plumwood means as a revised view of ecofeminist epistemologies (47). Works of ecofeminists focus on the domination of women and nature. But ecofeminism also demands the disintegration of dualism as the *eco* part of ecofeminism aims to create a body of ecologically informed corpus of critical study.

Ecocriticism postulates a transformative discourse and hence engages with the complex negotiations of nature and culture. Greg Garrard explains that the widest definition of ecocriticism is the "study of the relationship of human and the nonhuman, throughout human cultural history" (5). In the era of ecological crisis, the culture-nature divide is a conceptual prison and hence is detrimental to resolving environmental problems and our dealings with nature. Dualism prioritizes human cultural elements and imposes conceptual conditions on nature as a separate and alienated sphere. As in the words of Collingwood: "how can mind have any connection with something utterly alien to itself, something essentially mechanical, non-mental, namely nature?" (Collingwood 7). Ecocritics develop anti-dualist ways drawing on different philosophical and scientific discourses. Richard Kerridge states that ecocriticism from the beginning emphasizes the need to depart from the Cartesian tradition of dualism. Quoting Plumwood, Kerridge argues that environmentalists need to take the most important intellectual step—the break from this dualistic tradition: "developing environmental culture involves a systematic resolution of the culture/nature and reason/nature dualisms that split mind from body, reason from emotion, across their many domains of cultural influence" (366). In her "A Cyborg Manifesto" (1985), Donna Haraway subverted the categories of culture and nature, human and nonhuman, female and male. She uses the term "naturecultures" (2003) to explain that humans and natural environments (including nonhumans and more-than-humans) are intimately entangled in different places. Ynestra King's "transformative feminism" emphasizes the need "to pose a rational re-enchantment that brings together spiritual and material, being and knowing. This is the promise of ecological feminism" (202). King's *transformative feminism*, Merchant's *partnership ethics*, Warren's *ethic of care*, and

Plumwood's *ecosocial feminism* are critiques of dualism and hierarchical thinking and present a *shared perspective* that nature always cohabits with culture harmoniously in *contextually-specified complexes*. Our focus is somewhat similar: we consider bioregionalism as a complex of nature and culture. Therefore, bioregionalism as a multistranded eco-concept may intersect with the *shared perspective* and transdisciplinary voices decomposing the culture/nature dichotomy.

BIOREGIONALISM, BIOCULTURAL REGION AND THE CULTURE-NATURE NEGOTIATION

Our main thesis concerning bioregionalism and culture/nature dualism is that the biocultural region perspective allows for disintegrating the dualism on the level of context-specific analysis. What is the theoretical perspective of the biocultural region, *the possible mix to re-read environmental literature*? How does the biocultural region perspective converge the seemingly separated domains of culture and nature? It is fruitful to study what bioregionalism is and how dualism is disintegrated in the context of bioregionalism.

Bioregionalism coalesced as part of the development of the environmental movement and ecocritical study in the 1970s. The word "bioregion" was coined by Allen Van Newkirk (1975) in his writings: "biogeographically interpreted culture areas . . . called bioregions" (Aberley 22). The major bioregional thinkers are Peter Berg, Raymond Dasmann, Gary Snyder, Doug Aberley, Kirkpatrick Sale, Jim Dodge, and Stephanie Mills. Berg and Dashmann state that a bioregion refers to both the geographical terrain and a terrain of consciousness of a place and the ideas that have developed about how to live in that place ("Reinhabiting California" 36). Kirkpatrick Sale in *Dwellers of Land* (1985) presents a comprehensive study of bioregionalism. His view resonates with Berg and Dashmann that people determine the final boundaries of bioregions following their own sensibilities (43).

Jim Dodge explains that the term *bioregionalism* is formed from Greek *bios* (life) and French *region* (itself from Latin *regia* (territory); thus, the meaning is *life territory* or *place of life* (Taylor 67). Robert L. Thayer Jr. states that a bioregion is:

> literally and etymologically a "life-place"—a unique region definable by natural (rather than political) boundaries with a geographic, climatic, hydrological, and ecological character capable of supporting unique human and nonhuman living communities. Bioregions can be variously defined by the geography of watersheds, similar plant and animal ecosystems, and related, identifiable landforms

> (e.g., particular mountain ranges, prairies, or coastal zones) and by the unique human cultures that grow from natural limits and potentials of the region. (3)

Michael Vincent McGinnis traces the idea of bioregionalism in the primal native inhabitants or indigenous people who practised many of its tenets. Elaborating on the concept, McGinnis further states that bioregionalists stress the importance of reinhabiting one's place and earthly home in the modern social segregation of nature-society. A particular bioregion represents the intersection of "vernacular culture, place-based behaviour, and community" (3). This echoes Berg's stand on the concept of *reinhabiting* with the optimum object of living sustainably. Berg's bioregional concepts bordering on sustainability and interdependency and supporting other species are important steps to limit the modernist nature-society separation. Thus, it refers to the mutual and connected existence of all life forms on earth, creating sustainable human societies in unison with the natural world and flourishing of all species in a particular region.

John Charles Ryan quotes Wendell Berry's description of bioregionalism as "a local life aware of itself" (82). However, Ryan prefers to describe bioregionalism as "a local life aware of itself in its natural setting" (82). Place evolves from the interplay between humans and the environment, with the environment as a point of reference. Bioregionalism's emphasis on natural places is viewed as a possible solution to the recurring pattern of human negligence toward the natural world (84). Wendell Berry has argued that without possessing a complex knowledge about one's place and faithfulness to one's place on which such knowledge depends, it is inevitable that the place will be used carelessly and eventually destroyed (39). Knowledge about one's place or bioregional sensibility is a positive, proactive force to face the challenges of impending environmental crises and "learning to live-in-place" (Berg 6) sustainably. Ecocriticism and place studies share a common ground, specifically the literary reflections on place. Nancy Easterlin points out that place studies emphasize the importance of social attachments and human feelings about environmental perceptions of place. She states that ecocritics also aim to study how such a field might contribute to an enlightened view of the nonhuman natural world (232). In the works of Ursula K. Heise (2008) and Lawrence Buell (2005), the place remains important to the ecocritical investigation of diverse literature. Buell defines place as a space that is bounded as humanely meaningful through personal attachment, social relations, and physiographic distinctiveness (145). Therefore, bioregionalism and place studies explore place-based thinking, beginning with a particular region and place and ending with culture.

Culture and nature are distinguished from each other as two separate realms of reality (Haila 155). In their works, Yrjö Haila (2000), Plumwood and Anne

Elvey (2006) have studied that it is "possible to construct context-specific alternatives demonstrating that culture and nature belong together" (Haila 156). It is important to mention Plumwood's concept of the *cultural landscape*. Plumwood explains that the term cultural landscape disrupts the dualistic concept of the human (culture) being set apart from the natural world (nature). She details that there is a compatibility between indigenous (cultural) agency and nonhuman (natural) agency. The term cultural landscape acknowledges indigenous agency in the land (120–121). She observes that the concept of the cultural landscape cannot totally reject dualism, as both terms (culture and landscape) raise problems. Culture can be intended as human-created or human-influenced and landscape as a human product. The landscape often refers to an androcentric model that frames the land as passive and visually captured (122–123). Alison Byerly resonates with similar thoughts as she describes the aesthetic of the picturesque in the context of the National Park system in America. She points out that the American wilderness has been transformed from a sublime landscape into a series of picturesque scenes. The specific qualities of this picturesque aesthetic were based on principles derived from painting not from nature (53). Hence, this way of constructing the concept borders more towards dualistic thinking humans (culture) as the privileged source of contrast and land (nature) as the passive and subdued sphere.

The biocultural region attempts to depart from this dualism, provides an alternative perspective, and counters the limitations of the culture/ nature binary. When we study a bioregion, we consider the natural characteristics of a place, including its people. The biocultural region presents a convergence more comprehensively: the partnership of human culture with the place/ region. The perspective offers a human (re)connection with the land in the constantly changing world. This perspective presents an eclectic mix of thoughts engaging different schools of ecocriticism and place studies. Given this background, we may say that a biocultural region means a bioregion (a specific, local geographic area) in which human cultural relationship to the natural world develops and sustains to maintain the long-term functioning of the ecosystem they inhabit (or cohabit with biotic and abiotic communities).

Advancing a groundwork for an alternate and effective perspective, our study also outlines a few limitations that offer interesting lines of inquiry for further studies. The concept of the biocultural region is *context-specific* to bioregionalism, place studies, and ecocriticism. It is not *totalizing,* as it is impossible to eliminate the all-encompassing dualism completely. The perspective presents bioregion-specific interaction and interconnection between human culture and the ecosystem. It may give rise to blind spots in the human-nonhuman/animal interface in a particular bioregion. The culture/ nature dualism conceives that humans are associated with culture, whereas nonhumans exist in nature, insisting on two radically separate spheres. Given this assumed segregation, the concept of inclusion and ecological existence is contested (humans are superior to nonhumans because of their rationality and dominate other nonhuman species and abiotic communities). This thesis offers an interesting line of study to explore the interaction between culture-nature and human-nonhuman beyond segregation (exploring the intersection of ecocriticism, Animal Studies, and Human-Animal Studies).

BIOCULTURAL REGION: RE-READING ENVIRONMENTAL NONFICTION

Creative nonfiction on the environment, such as William Bartram's *Travels* (1791), Henry David Thoreau's *Walden* (1854), Aldo Leopold's *A Sound County Almanac* (1949), Edward Abbey's *Desert Solitaire* (1968), and Annie Dillard's *Pilgrim at Tinker Creek* (1974), discuss different bioregions in diverse ways. They all perceive the specific geographical areas through an increased awareness of human cultural practices. Bartram's *Travels* documents his extensive and passionate exploration of the southern frontier of the American landscapes, from California, and South Carolina to Choctaw and Cherokee lands and swamplands of the New World. His work is an interesting text to be studied through a biocultural region perspective as permanent records of Native American culture, the first colonists, and the mystery of the pristine landscape. Robert Sayre, in his article "William Bartram and Environmentalism" (2015), states that it is important to understand the essence of Bartram's vision in the North American setting. The first colonists found what they saw as "wilderness"—lands sparsely populated by peoples whose imprint on them was light—but the "civilizing" of them was rapid at the hands of the newcomers. He continues that this early capitalist civilization was agriculture-based, so the colonists maintained the expansionist logic while acquiring land. Bartram, during his trip, encountered several manifestations of this process (69). In *Pilgrim at Tinker Creek*, Annie Dillard places her voice in the forefront as she writes about the place. She presents the "American Indian culture a way of responding to nature that seems close to what she seeks" (Norwood 341).

Thoreau's deliberate act of living at Walden Pond in March 1845 exemplifies his *bioregional living*. Lawrence Buell, in "Thoreau and the Natural Environment" (1995), considers Thoreau as the "first American environmentalist saint" whose first intellectual prompting to study and write about nature was from books and literary mentors like Ralph Waldo Emerson. His career in pursuit of nature has become one of fitful, irregular, experimental, although increasingly purposeful, self-education in reading the landscape and pondering the significance of what he found there (171). Similarly, Leopold's *A Sand County Almanac* documents a particular bioregion—Wisconsin. Leopold writes about the vast functional ecosystem of Prairie and the human cultures organizing themselves in the grasslands. Edward Abbey's *Desert Solitaire* is about Arches National Park, a bioregion fostering a sense of place and connection with human culture. Scott Slovic, in his work, *Seeking Awareness in American Nature Writing: Henry Thoreau, Annie Dillard, Edward Abbey, Wendell Berry, Barry Lopez* (1992), writes that Edward Abbey's *Desert Solitaire* contains many examples of the harshness and unfamiliarity of the desert landscape. Slovic presents the contrasting ways of looking at *Desert Solitaire* and *The Monkey Wrench Gang*:

> Edwin Way Teale called *Desert Solitaire* "a voice crying in the wilderness, *for* the wilderness" ("Wild Scene," 7). And Grace Lichtenstein, reviewing *The Monkey Wrench Gang* for the *New York Times,* notes that Abbey "has been the most eloquent spokesman for angry nature-lovers . . . His message, that only a radical change in the American life-style or even more radical action will preserve the land for future generations has become a watchword among the growing minority of those who call themselves eco-freaks." (Slovic 102)

CONCLUSION

In the current era of environmental crisis, the planet is open to rampant human exploitation of nature. Dualistic thinking represents nature as passive and an object of exploitation, dominated and controlled by humans. As Cheryll Glotfelty points out, ecocriticism emphasizes that this ecological crisis is our making. She quotes Donald Worster that the global problem we face is not because of how the ecosystem functions but how our ethical system functions. We need to understand our impact on nature as precisely as possible to get through the crisis (xxi). Understanding anthropogenic impact on nature begins with the place or region we inhabit: the reciprocal relationship between humans and the place (*learning to live-in-place* sustainably). Understanding and acting are to go beyond the dualistic thoughts (culture/nature divide) and develop an alternate vision in which nature is co-evolving with humanity to save the planet from any anthropogenic calamity.

Connecting bioregionalism and environmental crisis, we may say that bioregional thinking and living (local to global, connections between place-based knowledge and global environmental change) is what the advocates of place study and ecocritical theorists strive to achieve. Bioregionalism seeks to address ecological crises on the local level. But the present urgency is to create a more globally sustainable world to solve anthropogenic threats. The key idea of bioregional thinking in the Anthropocene era is "Think globally, act locally" (Buell et al. 421). Since bioregions are shaped by characteristics of a particular ecosystem determined by social and cultural factors, it gives a scope to the coevolution of both human and natural systems. Therefore, environmental problems can be addressed locally, considering the ecological, social, cultural, and political constraints. Such practices will prove conducive to global sustainability.

Drawing on the works of bioregionalists and ecocritics, our study has shown that bioregionalism as an eco-concept presents "an alternative paradigm" (Sale 29) promoting a dialectical relationship between nature and culture. The eco-concept demands a rethinking of the current theoretical perspectives and proposes a groundwork for decomposing culture/nature dualism. The perspective of the biocultural region presents a critical reflection on this basic premise of the eco-concept. It focuses on human partnerships with a particular bioregion: the relationship between human culture and the ecosystem they cohabit with biotic and abiotic communities. The perspective is optimistic and works towards mutualism (culture-nature co-evaluation in the context-specific analysis of a bioregion).

REFERENCES

Abbey, Edward. *Desert Solitaire.* 1968. Simon and Schuster, 1990.

Aberley, Doug. "Interpreting Bioregionalism." *Bioregionalism.* Edited by Michael Vincent McGinnis. Routledge, 1998.

Bartram, William. *Travels.* 1791. Dublin: For J. Moore, W. Jones, R. McAllister, and J. Rice. 1793.

Berry, Wendell. "The Agrarian Standard." *The Essential Agrarian Reader: The Future of Culture, Community, and the Land.* Edited by Norman Wirzba. The University Press of Kentucky, 2003, pp. 23–33.

Buell, Lawrence. *The Environmental Imagination: Thoreau, Nature Writing, and the Formation of American Culture.* Belknap Press of Harvard University Press,1995.

———. *The Future of Environmental Criticism: Environmental Crisis and Literary Imagination.* Blackwell Publishing, 2005.

Buell, Lawrence, Ursula K. Heise, and Karen Thornber. "Literature and Environment." *Annual Reviews*, vol. 36, 2011, Pp. 417–440.

Bunting, Robert. *The Pacific Raincoast: Environment and Culture in an American Eden, 1778–1900.* University Press of Kansas, 1997.

Byerly, Alison. "Uses of Landscape." *The Ecocriticism Reader: Landmarks in Literary Ecology.* Edited by Glotfelty, Cheryll, and Harold Fromm. University of Georgia Press, 1996.

Collingwood, Robin George. *The Idea of Nature.* Clarendon Press, 1945.

Dillard, Annie. *Pilgrim at Tinker Creek.* HarperCollins, 2009.

Easterlin, Nancy. "Ecocriticism, Place Studies, and Colm Tóibín's 'A Long Winter': A Biocultural Perspective." *Handbook of Ecocriticism and Cultural Ecology*. Edited by Hubert Zapf. De Gruyter, 2016, pp. 226–248.

Gaard, Greta, editor. *Ecofeminism: Women, Animals, Nature.* Temple University Press, 1993.

Garrard, Greg. *Ecocriticism.* Routledge, 2004.

Glotfelty, Cheryll, and Harold Fromm, editors. *The Ecocriticism Reader: Landmarks in Literary Ecology*. University of Georgia Press, 1996.

Glotfelty, Cheryll, and Eve Quesnel, editors. *The Biosphere and the Bioregion: Essential Writings of Peter Berg*. Routledge, 2015.

Haila, Yrjö. "Beyond Nature-Culture Dualism." *Biology and Philosophy*, vol. 15, 2000, 155–175.

Haraway, Donna. *The Companion Species Manifesto: Dogs, People, and Significant Otherness*. Prickly Paradigm Press, 2003.

Heise, Ursula K. *Sense of Place and Sense of Planet: The Environmental Imagination of the Global.* Oxford University Press, 2008.

Kerridge, Richard. "Ecocritical Approaches to Literary Form and Genre: Urgency, Depth, Provisionality, Temporality." *The Oxford Handbook of Ecocriticism.* Edited by Greg Garrard. Oxford Academic, 2014.

King, Ynestra. "Feminism and the Revolt of Nature." *Heresies #13: Feminism and Ecology*, vol. 4, no. 1, 1981, pp. 12–16.

Leopold. Aldo. *A Sand County Almanac: With Essays on Conservation from Round River*. 1949. Ballantine, 1970.

McGinnis, Michael Vincent. "A Rehearsal to Bioregionalism." *Bioregionalism*. Edited by Michael Vincent McGinnis. Routledge, 1998.

Merchant, Carolyn. *Earthcare: Women and the Environment*. Taylor and Francis, 2014.

———. *Radical Ecology*. Routledge, 2005.

Norwood, Vera L. "Heroines of Nature: Four Women Respond to the American Landscape." *The Ecocriticism Reader: Landmarks in Literary Ecology*. Edited by Cheryll Glotfelty and Harold Fromm. University of Georgia Press, 1996.

Plant, Judith. "Searching for Common Ground: Ecofeminism and Bioregionalism." *Living with Contradictions: Controversies in Feminist Social Ethics*. Edited by Alison M. Jaggar. Westview Press, 1994, pp. 672–676.

Plumwood, Val. "Ecosocial Feminism as a General Theory of Oppression." *Key Concepts in Critical Theory: Ecology.* Edited by Carolyn Merchant. Humanities Press, 1994, pp. 207–219.

———. *Feminism and the Mastery of Nature*. Routledge, 1993.

———. "The Concept of a Cultural Landscape: Nature, Culture and Agency in the Land." *Ethics and Environment*, vol. 11, no. 2, Fall–Winter, 2006, pp. 115–150.

Ryan, John Charles. "Humanity's Bioregional Places: Linking Space, Aesthetics, and the Ethics of Reinhabitation." Humanities, vol. 1, 2012, pp. 80–103. https://doi.org/10.3390/h1010080.

Sale, Kirkpatrick. *Dwellers in the Land: The Bioregional Vision.* Sierra Club Books, 1985.

Sayre, Robert. "William Bartram and Environmentalism." *American Studies*, vol. 54, no. 1, 2015, pp. 67–87. http://www.jstor.org/stable/24589488.

Slovic, Scott. *Seeking Awareness in American Nature Writing: Henry Thoreau, Annie Dillard, Edward Abbey, Wendell Berry, Barry Lopez*. University of Utah Press, 1992.

Taylor, Bron. "Bioregionalism: An Ethics of Loyalty to Place." *Landscape Journal*, vol. 19, no. 1 and 2, 2000, pp. 50–72.

Thayer, Robert L. *Life Place: Bioregional Thought and Practice*. University of California Press, 2003.

Thoreau, Henry David. *Walden*. 1854. *Walden and Resistance to Civil Government*. Ed. William Rossi. Norton Critical Ed. 1992.

Warren, Karen J., editor. *Ecofeminism: Women, Culture, Nature*. Indiana University Press, 1997.

———. *Ecofeminist Philosophy: A Western Perspective on What It Is and Why It Matters*. Rowman & Littlefield Publishers, 2000.

Chapter 7

Reading Ecophobia in the Capitalocene

Brian Deyo

In a remarkable scene from J.M. Coetzee's *Dusklands*, the first-person narrator, Jacobus Coetzee, an eighteenth-century Dutch settler, frontiersman—and self-described "tamer of the wild"; or, alternatively, "destroyer of wilderness"—writes a retrospective account of his encounters with the Indigenous peoples of Southern Africa (Coetzee 77, 79). With the coolly clinical, expertly detached tone of an ethnographer, Jacobus recounts what he sees upon arriving at a Khoikhoi camp. After imparting a dry account of the "forty huts arranged in a rough circle," Coetzee notes how the "apex" of each "hut" is "open," affording "the Hottenntot abed a barred view of the night sky," after which he briskly infers—and with astonishing cocksurety—that the opening onto the cosmos "has led to neither a special relationship with the sky gods," "nor a Hottentot astrology"; rather, it is, as he declares, "*nothing* but a smoke hole" (71, my emphasis). The officiousness of this and similar pronouncements are made with such regularity and zeal as to imply his self-image as a superior, godlike being is utterly dependent upon his representations of Indigenous peoples as utterly void of value: hence, his inveterate imputations of unregenerate animality to them are understood to support an idealized humanity the reader comes to associate with Cartesian rationality. As David Attwell contends, Jacobus's stigmatization of Indigenous peoples is one of the "principal means" through which he "achieve[s] self-affirmation"—which would also appear to be contingent on his capacities for self-mastery and control over others, including the natural world. Atwell cogently aligns Jacobus's metaphysical disquisitions on the "nature" of Indigenous peoples, societies, and cultures—not to mention the "nature" of nature, suggestively conflated with the former—with "scientific discourses that . . . evolved in the wake of the Enlightenment," the selfsame discourses that instantiate and support "the narrator's subject-position": or, what Coetzee's parodic re-enactment of settler colonial histories satirically diagnoses as a quite lame, albeit revealingly unsettling "fiction of self-assertion" (Attwell 39, 48). Employing a postcolonial-ecocritical framework to coordinate Jacobus's "derogatory representations" of nature and Indigenous peoples with ecophobia—provocatively defined by Simon Estok a little over a decade ago as "the contempt and fear we feel for the agency of the natural environment"—I've argued that Coetzee's literary excavations of the past in *Dusklands* limn the dark ecophobic core of settler colonial societies, "delineating the tragic consequences" of the entrenchment of a form of subjectivity that is tellingly obsessed with the mastery and domestication of "otherness," both without and within: what the ecofeminist philosopher Val Plumwood variously refers to as "the master" or "dominator identity" of western culture in her landmark study, *Feminism and the Mastery of Nature* (Deyo 94; "Theorizing" 207; Deyo 99; Plumwood 5).

I begin this entry on ecophobia with *Dusklands* for a variety of reasons. For starters, the novella provides a truly breathtaking range of opportunities for "theorizing ecophobia," in the parlance of Estok, "through ecocritical lenses" (*Ecocriticism and Shakespeare* 5). For Estok, theorizing ecophobia is tantamount to the forms and cultural practices of "reading" unique to the discipline of literary studies, as well as the critical traditions of thought associated with the humanities more generally: to elaborate, when "we're" theorizing ecophobia via literary texts, "we're," in a manner of speaking, reading (or thinking critically about) language, culture and history, not to mention the forms of subjectivity constituted by the politics of power; in other words, when "we're" reading for ecophobia, "we're" reading "ourselves," albeit with a sober, clear-eyed attentiveness to "our" relations with (and attitudes toward) nonhuman nature, the world(s) of animals, plants, microbes and the elements, indeed the cosmos itself. Following Cheryll Glotfelty's succinct definition of ecocriticism as "the study of the relationship between literature and the *physical* environment," one can argue that, inasmuch as any literary text, historical or contemporary, has the potential to illuminate *homo sapiens'* varied relations with the more-than-human world, it is of the utmost importance to build and evolve the educational infrastructure of the field, share its major concepts and insights, both internationally and across a wide variety of social institutions—primary and secondary education, colleges and universities, health care, government—and with a singularly stringent focus on the twin existential crises of climate breakdown and the sixth extinction (Glotfelty xviii, my emphasis). The fact this hasn't happened—despite the robust, detailed, and varied scientific evidence we have of the planet's unraveling—says something about us, which is why "literary studies scholars," says Iris Ralph, would do well to actively promote scholarship and pedagogies that critically examine the "ecophobic content" of literary texts (Ralph 402). Ralph rightly contends that this would be an opportune and "obvious strategy for making" terms such as "'ecophobia and its like (for example, 'speciesism' and 'ecocide') household words in schools, offices, airports, gas stations, corner stores, food courts, and shopping malls" (402).

Nevertheless, as she notes, there's something about the term that generates resistance—perhaps ecophobia—even amongst prominent ecocritics. Citing the extent to which the concept of ecophobia has "gained currency" amongst ecocritics since the publication of Estok's "Theorizing in a Space of Ambivalent Openness: Ecocriticism and Ecophobia," Ralph also points out that "discussion of ecophobia does not appear in several recent and distinguished anthologies of ecocriticism": an important, if disconcerting, reminder of the social, psychological, rhetorical and political challenges that ecocritics invested in the project of mainstreaming its meanings, uses, and applications beyond the groves of academe will certainly face in the fight for climate justice, now, until 2050, and beyond (401). Perhaps the ambivalence Ralph ascribes to many ecocritics—including several of its leading practitioners—testifies to the systemic entrenchment (or hegemony) of anthropocentric ideas and priorities within modern societies and cultures, signifying a collective and quite tragic failure of the imagination on the part of humanity, at least a critical portion of us. As Raj Patel, Jason W. Moore and the practitioners of "world-ecology" helpfully remind us, "[we] may all be in the same boat when it comes to climate change, but most of us are in steerage" (Patel and Moore 24). The authors—like Melanie Klein, Camille Dungy, Kathleen Jamie, and (if we need reminding) countless others—are keen to clarify that the planetary catastrophe conjured by words like the Anthropocene is largely the consequence of the ideas and material practices of "*some* humans," namely "those in charge of conquering and commercializing a world that counts only as dollars" (24). As the above propositions clearly indicate, Marxian class analyses are indispensable to a clear understanding of the origins of environmental "problems," like climate breakdown and the sixth extinction. Patel and Moore maintain, over and against commonplace correlations of environmental crisis with "natural human behaviors" assumed to have existed in "humans" from time immemorial, that capitalism—namely, a "rather specific interaction between humans and the biological and physical world"—"has brought us to this point" (5). Thus, if we are to understand the roots of our predicament, a collective understanding of the manifold ways in which capitalism has shaped and continues to shape the "web of life"—and vice versa—is imperative (22–24). But, as our collective knowledge of, for example, the science of climate has grown apace just over the past decade—and without galvanizing the force of collective action required to meaningfully address the crisis—it's becoming quite clear to many that we need poetry, in the broadest sense of the word, to be able to feelingly envision what rational historical analysis helps us to see.

Frederic Jameson is said to have asserted in conversation that "it's easier to imagine an end of the world than an end to capitalism"; similarly, one might posit that it's easier to imagine *homo sapiens* as essentially self-interested beings, driven more by greed, competition, acquisitiveness, lust for power (and the like) than cooperation, fellow feeling, love, and an enduring desire for beauty—or, articulated a touch differently, a vital and fulfilling sense of connection with the more-than-human world. I mention this simply to underscore the sheer force and appeal of capitalist ideology, which, lest we forget, shapes the unconscious and conscious desires of even the most vigilant, self-critical of ecocritics; secondly, as Jameson's diction in the above formulation suggests, the planetary catastrophe unleashed by colonialist capitalism since 1492 is a problem of, by, and for the imagination, the capacity for which might be said to be a defining (if not entirely distinctive) feature of an animal that has only very lately come to define itself as "human," if only to differentiate itself from the rest of nature to better exploit "it"—and for the sake of an idea, hoisted up on high as an idol for the powerful and fortunate few who've held sway over this planet's destiny . . . and for a proverbial blink of an eye, at least if imagined against the prehistories and histories of this particular species' highly improbable, painfully brief, although singularly wondrous tenure here. In sum, for Jameson—as it is for us—the imagination is both problem and solution, which is why this short entry on ecophobia argues that the cultural practice of reading ecophobia through the histories and literatures of the Capitalocene is a matter of life and death; or, to elaborate, of more-than-human flourishing and mass extinction–not to put too fine a point on what *ought* to be obvious. Concerning the question as to why it isn't, Melanie Challenger posits that the evolution of human consciousness and language, coupled with a capacity for culture and story—which is to say, the ability to make meaning—produces the basis for the denial of what should register as common sense, at least from a strictly rational point of view; and, whether it's climate denialism, Holocaust denialism, or the denial that capitalism is destroying the ground of our existence, this feature of our being is ultimately grounded in a denial of our animal condition. "In truth," she says, "we live inside a paradox: it's blindingly obvious that we're animals and yet some part of us doesn't believe it" (Challenger 9). As I've stated before in an article that employs psychoanalytic and existential analytics to explore the cultural and historical dynamics of ecophobia—with special attention to the legacies of Enlightenment discourse on the "the question of the animal"—the material conditions of modernity, as well as the ideas of "progress" generated by the political economies of colonialist capitalism, engender what we could imagine (and thereby conceptualize) as a collective alienation from our animality—and, by logical implication, the basis of our kinship and solidarity with the more-than-human-world. In other words, the world-picture offered by modernity—as if somehow ontologically predisposed to the needs and desires of an abstract, historically contingent, and therefore imaginatively circumscribed "humanity"—is as monumentally productive for the West (and its governing elites) as it is costly for the vast majority of humans, not to mention the ecological systems from which they evolved and depend:

> [I]f culture enables humans to efficiently mobilize their energies to exploit nature, does it not seem reasonable to assume anthropocentric cultures that aggressively promote ecophobic impulses, attitudes, and dispositions would be able to exploit nature with greater efficiency, thus enabling them to become hegemonic—even if for limited periods of time? This should be common sense. Tragically, it isn't, which is why the concept of ecophobia ought to be a foundational feature in the practice of global environmental education. ("Denial of Death" 446–447)

So, how can we begin to think of ways to redress a picture of the world that has led to the systematic institutionalization of anthropocentric ideas—reflected in the material and technological infrastructures of petro-capitalism that ensnare us—and on a seemingly unimaginable scale? And if this latest instantiation of modernity—upon which we depend to carry out and sustain our very vulnerable (and increasingly uncertain) day-to-day lives—is widely perceived to be immovable or inevitable, what hope then? What gets in the way of the imagination's capacity to recognize what kind of beings we are, where we are, and how we are, together, right now, amidst this singular—and singularly urgent—concatenation of circumstances, signified by words, such as the Anthropocene, climate breakdown, the sixth extinction, too-late-capitalism, and the like? And, finally, how might "ecophobia studies," which has lately "emerged as a scholarly field of inquiry," help us to attain greater insight and clarity regarding our role, status, place, and agency vis-à-vis the more-than-human world? (*Ecophobia Hypothesis* 58).

These challenging and important questions are especially pertinent in the context of a twenty-first century media ecology that is drenched with anthropocentric ideology—what Estok broadly conceptualizes as cultures bearing the deep imprint of "ecophobic human exceptionalism" (*Ecophobia Hypothesis*, 25). The renowned journalist and climate activist, Naomi Klein, proposes that we widen our perspective and consider the impressive panoply of human societies and cultures, historical and contemporary, that are not organized around assumptions of fixed ontological differences between humans and nature. In an essay that employs the postcolonial concept of "othering" to explain the concept of the "sacrifice zone"—the idea that "you can't have a system built on sacrificial places and sacrificial people unless intellectual theories that justify their sacrifice exist and persist"—she says that "[w]e often hear climate change blamed on 'human nature,' on the inherent greed and short-sightedness of the species" (Klein). Critiquing the logics of such defeatist and cynical explanations of ecological crisis—particularly the facile and unhistorical ways in which the complexity and diversity of human nature are "essentialized to the traits" that created it—Klein cogently argues that such diagnoses ought to be seen for what they are: ideologically driven pseudo-explanations that let the very "systems" that have created the climate crisis, like "[c]apitalism, colonialism, and patriarchy," completely "off the hook," thereby "eras[ing] the very existence of human systems that organized life differently." Klein's assessment of the origins of the climate crisis—and socio-ecological crisis more generally—strikingly resonates with contemporary critical theories, like critical posthumanism, that are committed to the work of "deconstructing" discourses and concepts (like, for example, the Anthropocene) that reify and reproduce, unwittingly or not, Enlightenment conceptions of a universal humanity. Much of this work hews to the idea that the "human"—or what Stefan Herbrechter refers to as the subject of liberal humanism—is something along the lines of a fetish, what Moore and Patel liken to a practical instrument of bourgeois rule over racialized and naturalized (or animalized) others since 1492.

While the work of interrogating and demystifying bourgeois, anthropocentric discourse promises to institute cultural bases for rethinking, rewriting, and/or reimaging what it is or might mean to be human at this unique historical conjuncture—perhaps opening us up to what might amount to a reclaiming of a more ecologically grounded, empirically warranted, and therefore less ecophobic consensus concerning our common humanity—it's not enough to simply assume that, because contemporary forms of subjectivity are mediated by cultures of ecophobic human exceptionalism, an international project of cultural renovation at scale would somehow put paid to ecophobia, once and for all. Estok agrees with contemporary environmental thinkers, like Klein, who "theorize that the links between environmental destruction and capitalism cannot be ignored"; however, he's also right to point out that "capitalism is certainly not the *cause* of our ongoing environmental problems": "ecophobia," similar to social maladies such as, for example, "sexism, heterosexism, and racism," "predate capitalism by millennia" (*Ecophobia Hypothesis* 64, my emphasis). In other words, although ecocritics, cultural critics, and critical theorists have a hugely important role to play in the education of broad and diverse publics on the manifold ways anthropocentric cultural systems reinforce ecophobic tendencies fueling the immensely destructive capacities of colonialist capitalism, Estok is careful to acknowledge that at least a "certain amount of our ecophobia" is based in genetics (*Ecophobia Hypothesis* 25). Culture *and* biology, including our species' rich and complex evolutionary history, not to mention our varied entanglements with the critters, plants, and microbes that constitute ecological systems needs consideration—assuming, of course, that our aim is to understand ecophobia well enough to manage it and mitigate the harm we inflict on the more-than-human world, as well as each other.

This is why we might want to imagine ecophobia as an exceedingly complicated phenomenon that, like climate or capitalism—or "ourselves," for that matter—cannot be adequately understood from the vantage point of one, but rather a vast ensemble of disciplinary perspectives: like, for example, the work currently underway in the interdisciplinary environmental humanities, which bridges work in the humanities, the social sciences, and the natural sciences to deepen our understanding of the roots of socio-ecological crisis, and thereby take meaningful collective action to envision and build viable and more just futures. It may also be helpful to think of ecophobia, given its connections with biological and cultural evolution, as a "wicked problem," or even a hyperobject—a phenomenon, being, or process so "massively distributed in time relative to humans" as to defy cognition (Morton 1). Perhaps this is just a roundabout way of saying that ecophobia is a problem we're probably stuck with, something we need be prepared to live with—to the best of our collective capacities, and with a salutary measure of humility to boot. In fact, one can argue that the study of ecophobia has great potential on many counts, particularly concerning the degree to which it, at least in theory, helps us to *cultivate* humility, an affective disposition toward the more-than-human world that might be said to precondition other socially and ethically productive affects—such as awe, wonder, sorrow, grief, gratitude and, perhaps most vital, love.

This is why reading, studying, and listening to the stories of others is so crucial in the Capitalocene, especially the most vulnerable. It is often repeated, especially in activist communities, that we "know" the science of climate, and yet it's a conundrum as to why people don't seem motivated to meaningfully act on what they know. The social sciences are well positioned to guide us toward satisfying and insightful answers to this and similar questions. And as disciplines like social psychology expand our knowledge and understanding of the litany of variables and factors that affect our species' capacities for collective action to prevent the calamity of runaway climate change, we still have literature and the discipline of literary studies, not to mention an international network of colleges and universities that have prioritized the study of literature and culture to help us navigate the complexities of the present—a powerful sign that, lest we forget, we are creatures who necessarily yearn to create and connect with art, collectively, with each other. As Anna Kornbluh claims, this is so, precisely because "it is an exercise of our being" ("Responses"). From time immemorial, humans have turned to art and storytelling—in times of great difficulty and sadness, as well as profound gladness and joy—to make sense of who and what we are, to ask where we came from, where we are, where we're going, and how we should spend our short, uncertain, and therefore precious time together. I've argued elsewhere that the study of literature, particularly the genre of tragedy, "encourages a collective recognition of our shared, mortal condition with our animal cousins"—and that, inasmuch as this recognition is deeply and lastingly felt, we may use it to "enliven our capacities for sympathy and love, thereby honouring the evolutionary heritage with which our species is so richly endowed" ("Tragedy" 209). Reading, interpreting, and asking questions about ecophobia in the Capitalocene is to take the indispensable and utterly worthwhile risk of self-knowledge, which might just entail possibilities as endless as the universe—as well as the imagination, which the great visionary poet William Blake memorably, movingly, and beautifully equated with "Nature" itself (Blake).

REFERENCES

Abrams, David. *The Spell of the Sensuous: Perception and Language in a More-Than-Human World*. Pantheon Books, 1996.

Alexander, Bryan. *Universities on Fire: Higher Education in the Climate Crisis*. Johns Hopkins University Press, 2023.

Atwell, David. *J.M. Coetzee: South Africa and the Politics of Writing*. University of California Press, 1993.

Blake, William. *The Portable Blake*, edited by Alfred Kazin. Penguin, 1976.

Bridle, James. *Ways of Being: Animals, Plants, Machines: The Search for a Planetary Intelligence*. Farrar, Straus & Giroux, 2022.

Challenger, Melanie. *How to Be Animal: A New History of What It Means to Means to Be Human*. Penguin, 2021.

Coetzee, J.M. *Dusklands*. Penguin, 2017.

Dean, Jodi. "Capitalism is the End of the World." *Mediations: Journal of the Marxist Literary Group*, vol. 33, no. 1–2, 2019–2020, mediationsjournal.org/toc/realism-reevaluated. Accessed June 17, 2023.

Deyo, Brian. "Ecophobia, the Anthropocene, and the Denial of Death." *ISLE: Interdisciplinary Studies in Literature and the Environment*, vol. 26, no. 2, Spring 2019, pp. 442–455.

———. "Rewriting History/Animality in J.M. Coetzee's *Dusklands* and Richard Flanagan's *Wanting*." *ariel: A Review of International English Literature*, vol. 44, no. 4, 2013, pp. 89–116.

———. "Tragedy, Ecophobia, and Animality in the Anthropocene." *Affective Ecocriticism: Emotion, Embodiment, Environment*, edited by Kyle Bladow and Jennifer Ladino. University of Nebraska, 2018, pp. 195–212.
Estok, Simon. *Ecocriticism and Shakespeare: Reading Ecophobia*. Palgrave Macmillan, 2011.
———. *The Ecophobia Hypothesis*. Taylor & Francis, 2018.
———. "Theorizing in an Ambivalent Space of Openness: Ecocriticism and Ecophobia." *ISLE: Interdisciplinary Studies in Literature and the Environment*, vol. 16, no. 2, Spring 2009, pp. 203–225.
Fisher, Mark. *Capitalist Realism: Is There No Alternative?* John Hunt, 2022.
Freud, Sigmund. *Civilization and Its Discontents*. Translated and edited by James Strachey. Norton, 1961.
Glotfelty, Cheryll, and Harold Fromm, editors. *The Ecocriticism Reader: Landmarks in Literary Ecology*. University of Georgia Press, 1996.
Gumbs, Alexis Pauline. *DUB: Finding Ceremony*. Duke University Press, 2020.
Hartman, Steven and Patrick Degeorges. "DON'T PANIC: Fear and Acceptance in the Anthropocene." *ISLE: Interdisciplinary Studies in Literature and the Environment* vol. 26, no. 2, Spring 2019, pp. 456–472.
Herbrechter, Stefan. *Before Humanity: Posthumanism and Ancestrality*. Brill, 2021.
———. *Posthumanism: A Critical Analysis*. Bloomsbury, 2013.
Herbrechter, Stefan, interviewed by Will Brehm. "What Is Critical Posthumanist Education?" *FreshEd*, 123, podcast audio, July 16, 2019. freshedpodcast.com/stefanherbrechter/. Accessed June 17, 2023.
Klein, Naomi. "Let Them Drown: The Violence of Othering in a Warming World." *London Review of Books*, June 2, 2016, lrb.co.uk/the-paper/v38/n11/naomi-klein/let-them-drown. Accessed June 17, 2023.
Kornbluh, Anna. *Immediacy: Or, The Style of Too Late Capitalism*. Verso, 2024.
———. "Responses to Rita Felski's *Hooked: Art and Attachment*." NONSITE.ORG, April 2, 2021, https://nonsite.org/responses-to-hooked-art-and-attachment/#. Accessed June 17, 2023.
Krenak, Ailton. *Ideas to Postpone the End of the World*. House of Anansi, 2020.
Morton, Timothy. *Hyperobjects: Philosophy and Ecology after the End of the World*. University of Minnesota Press, 2013.
Patel, Raj, and Jason W. Moore. *A History of the World in Seven Cheap Things: A Guide to Capitalism, Nature, and the Future of the Planet*. Verso, 2018.
Plumwood, Val. *Feminism and the Mastery of Nature*. Routledge, 1993.
Ralph, Iris. "Ecophobia and the Porcelain Porcine Species." *ISLE: Interdisciplinary Studies in Literature and the Environment*, vol. 26, no. 2, Spring 2019, pp. 401–412.
Sharpe, Christina. *In the Wake: On Blackness and Being*. Duke University Press, 2016.
Slovic, Scott, Swarnathala Rangarajan, and Vidya Sarveswaran, editors. *Bloomsbury Handbook to the Medical-Environmental Humanities*. Bloomsbury, 2022.

Chapter 8

Eco-Deconstruction, or (Post)Humanism of the Other (Nonhuman)

Philippe Lynes

This chapter explains how eco-deconstruction might only be said to operate as an *eco-concept* insofar as it undoes and reroutes the concept of the concept itself. Just as Jacques Derrida says of his neologism *différance*—designating the spatial differentiation and temporal deferral anterior to, and constitutive of, every identity, every process of individuation (materio-physical, organic or politico-cultural)—I would wager that eco-deconstruction is *not* an eco-*concept*, nor can eco-deconstruction prescribe or be reduced to any *method* of environmentalist theorizing or praxis. Akin to Derrida's inaudible substitution of "*a*" for "*e*" in *différance*, "eco-deconstruction" supplementarily, even gratuitously re-iterates and ex-scribes on the outside of "d*eco*nstruction" the unheard "*eco-*" inside it, accentuating the deregulating mutations of the *oikoi* of ecology and economy already at play within it, interrupting the fluid circulation of whatever "concepts" might follow from their homeostatic, ideal, or proper functioning. This nonetheless has incalculably important implications for the environmental (post)humanities, if we understand these as caught an impossible *triple* bind, interpellated to adjudicate between what Thomas Van Dooren, Eben Kirksey and Ursula Münster outline as three allegedly *incommensurable* demands: "social justice in a humanist vein, ethics focused on the well-being of individual entities (usually nonhuman animals but to a lesser extent plants, fungi, stones, and others), and an environmental ethics concerned primarily with the health of ecosystems and species" (Van Dooren et al. 15). For the experience or trial of this incommensurability is also what nourishes and sustains the environmental humanities themselves; as Astrida Neimanis, Cecilia Åsberg and Johan Hedrén put it, "if we are to champion the need for environmental humanities today, it must be a version of this field that self-reflexively acknowledges *and even nurtures* its own contradictions, variances, and necessary open-endedness" (Neimanis et al. 69). And if any approach can be said to *nurture contradiction* and promote methodological *open-endedness*, it is deconstruction, which could even be said *to nature* contradiction as the very process of *nature denaturing itself.*

How, then, does eco-deconstruction motivate yet resist eco-conceptualization? In the second volume of his 1994–1996 seminar on hospitality, Derrida signals a madness at play within the very concept of hospitality, demonstrating the deconstructive derangement that *the concept itself* undergoes when confronted with the contradictions of hospitality: the contradictions of dwelling with others—nonhumans and humans—on earth:

> I say "indeed of the concept within hospitality," because the contradiction we are speaking of (atopian: madness, extravagance, in Greek: *atopos*) produces or registers this auto-deconstruction within every concept, within the concept of the concept: not only because hospitality undoes, should undo the grasp, the stranglehold (the *Begriff*, the *Begreifen*, the capture of *concipere*, *cum-capio*, *comprehendere*, the force or violence of "prehending" as "comprehending": hospitality is, *must* be, *owes it to itself* to be inconceivable and incomprehensible), but also because . . . every concept therein opens itself to its contrary, reproducing in advance, in the relation between one concept and another, the contradictory and deconstructive law of hospitality. Each concept becomes hospitable to its other, to an other of itself that is no longer even *its* other. (*Hospitalité II*, 148)

The madness by which any eco-concept should become hospitable to *an other* that is no longer *its* other, this chapter will show, is how I see eco-deconstruction intervening in the key debates at stake in the environmental humanities. In the philosophical logic that grounds the humanism of the humanities, a concept functions by generalization, always requiring another concept, an opposite from which to differentiate itself. But this generalization necessarily presupposes a hierarchizing reappropriation of the other. The concept of "culture" has that of "nature" as *its own* other, just as that of "the human" has "the animal" as *its own* other. Eco-deconstruction, by contrast, invites us to inquire into the site, or rather the *nonsite* or *nonphilosophical site* out of which this differentiation and reappropriation takes place and comes apart: it poses eco-philosophical questions from an *other site* that is no longer the other *of* philosophical conceptuality. As Derrida puts it, "my central question is: From what site or nonsite (*non-lieu*) can philosophy as such appear to itself as other than itself, so that it can interrogate and reflect upon itself in an original manner?" ("Deconstruction and the Other" 140–41)

The task of deconstruction, he adds, is "to discover the nonplace or *non-lieu* which would be the *other* of philosophy" (Ibid., 143–44). In opening up the concept to *an* other that would no longer be *its* other—that is, no longer an *intra-philosophical* or *intra-conceptual other*—by marking its inscription in a wider, ecological textuality, eco-deconstruction would thus constitute a *nonconcept*: "it remains nonconceptual; and because it has no oppositional or predicative generality, which would identify it as *this* rather than *that*, [it] cannot be defined within a system of logic—Aristotelian or dialectical—that is, within the logocentric system of philosophy" (Ibid., 143).

Building on Derrida's reading in *Hospitality, Volume II*, this chapter's title plays on that of Emmanuel Levinas's book, *Humanism of the Other*, whose translation of the French *Humanisme de l'autre homme*, of the other Man or Human, already complicates the field of our problematic—and not only because, as Derrida notes, "for non-fortuitous reasons, there is no concept of nature, no reference to a state of nature in Levinas, it seems to me, and this is of great consequence" (*Hospitalité II*, 70.) By teasing out these consequences, I wish to show how deconstruction lets us reinterpret living together well with other nonhumans and humans according to a logic of radical separation and non-relationality. I begin with an overview of the contemporary "environmental posthumanities," before turning to the ecocritical and environmental-philosophical roots of "eco-deconstruction," and conclude by demonstrating the insights shared by both in a brief reading of Derrida's *Hospitalité, Volume II.*

Rosi Braidotti's *Posthuman Knowledge* situates what she calls the posthuman condition at "the convergence of posthumanism on the one hand and post-anthropocentrism on the other, within an economy of advanced capitalism. . . . The former focuses on the critique of the Humanist ideal of 'Man' as the allegedly universal measure of all things, while the latter criticizes species hierarchy and anthropocentric exceptionalism" (Braidotti 2). The posthuman *convergence* thus occurs at the intersection of the Fourth Industrial Revolution, with its legacies of classist, colonial, racist and sexist violence against other humans, and the Sixth Mass Extinction of the planet's nonhuman species, "two parallel and to a certain extend specular forms of acceleration: the systemic accelerations of advanced capitalism and the great acceleration of climate change" (Ibid., 2). For Neimanis et al., the environmental humanities can resist these accelerations through a turn to a "critical posthumanities" (Neimanis et al. 90). Drawing from the feminist genealogies of posthumanism, this turn would resituate the "green" field as attentive to what Bruno Latour and Donna Haraway called "naturecultures," thereby opening up an ethics that would

> recognize the necessity of re-imagining the relationships between humans, other animals, waters, lands, and other "Earth Others" . . . as a cornerstone of continuing to live well with all these "others" on a changing planet. It would also, again, stress that differently situated humans (according to class, gender, age, geography, ethnicity, and so forth) will be affected very differently by environmental challenges. (Ibid., 83)

Now, as with ecofeminism, Derrida had long theorized the intersections of anthropocentrism, androcentrism, and the metaphysics of self-identical being or presence (logocentrism) through the schema of what he called *carno-phallogocentrism*:

> install[ing] the virile figure at the determinative centre of the subject. Authority and autonomy . . . are, through this schema, attributed to the man (*homo* and *vir*) rather than to the woman, and to the woman rather than to the animal. And of course to the adult rather than to the child. The virile strength of the adult male, the father, husband, or brother . . . belongs to the schema that dominates the concept of subject. The subject does not want just to master and possess nature actively. In our cultures, he accepts sacrifice and eats flesh. (*Points* 280–81)

By revealing the carno-phallogocentric structure of Western subjectivity, Derrida would come to side with a certain antihumanism, albeit not without reserve. As he paraphrases Levinas in the *Hospitality* seminar, "antihumanism is good insofar as it rids us of a satisfied, complacent human subjectivity . . . , one must even abandon the notion of the person, of subjectivity, because it is still a matter of an ego that is something because it is still a being, because it still is; so, one must go beyond being; exception to essence, exception to being; so, antihumanism has this virtue" (*Hospitalité II* 350). Deconstruction likewise importantly anticipated what Christine Daigle identifies as the target of the environmental posthumanities: "the human exceptionalism that grounds oppressive regimes which, in relation to the environment and nonhuman animals, justify extractive and destructive practices that have led to the Anthropocene and the 6th mass extinction event" (Daigle 881–82). Why, then, has deconstruction's recognition of this convergence between human and nonhuman oppression so often gone unnoticed in contemporary scholarship? Perhaps due to a misunderstanding of the complex nature of intersectionality and relationality that deconstruction deploys, but more likely due to a wider "theory-phobia" in academia.

Anti-humanist critiques of capitalism on the one hand, and anti-anthropocentric critiques of environmental degradation on the other, have indeed given rise to rather *divergent* practices of scholarly interpretation. Deconstruction, however, also allows us to articulate and interrogate otherwise this divergence, which lies at the heart of what Neimanis, Åsberg and Hedrén call the problem of "compartmentalization" in the environmental humanities. As they write,

> the obvious danger of compartmentalization is that the links between economy (e.g. growth), culture (e.g. consumerism) and environmental degradation and resource depletion will not be seriously explored. Another less obvious, but equally significant danger is that compartmentalization might *cover over human difference* (on the basis of gender, sexuality, race and ethnicity, age, bodily ability, geographical location, social and economic status) in the face of environmental challenges. (Neimanis et al. 79)

The compartmentalization of anti-humanism and post-anthropocentrism, of culture and nature, in other words, "risks failing to address the integrated nature of environmental injustice, across questions of racism and coloniality, gender and sexual difference, poverty, social exclusion and other ethical domains." (Ibid.) This disjunction is most readily visible in discourses concerning the "Anthropocene," which for Neimanis et al. "risk overwriting important differences among human populations and covering over uneven power distributions both in terms of responsibility for and vulnerability in the face of environmental problems" (Ibid. 68). Against this socio-political blindness, Daigle explains, newer discourses such as the Androcene, Anthrobscene, Capitalocene, Chthulucene, Plantationocene, Thanatocene, Thermocene, Trumpocene or White-Supremacy-Cene wish to make clear that

> identifying the whole species as responsible obscures that the *Anthropos* in question is a white, cisgendered, heterosexual, able-bodied, middle to upper class, man. Since it is mostly that being that is able to respond—that can exercise response-ability, an important concept for feminist posthumanists—failing to identify it by using the umbrella term "Anthropocene" misdirects the political agenda. (Daigle 892)

However, Daigle is careful to note that the way one approaches the Anthropocene and the 6th mass extinction differs

> whether one adopts a posthumanist new materialist stance or not. While environmental humanities scholars may be devising ways for humans to cope and thrive in a world that is facing multiple ecological—and many other types of—crises, making human well-being their central concern, environmental posthumanities scholars examine the concepts themselves and unearth the many human and humanist biases at work in understanding the problems we face. (Daigle 890)

Such post-anthropocentric approaches may have enjoyed a brief moment with what was called the *material turn* or *the nonhuman turn*, particularly in approaches grouped around new materialism (agential realism, feminist materialism, new vitalism, vital materialism, with authors including Stacy Alaimo, Karen Barad, Braidotti, Jane Bennett and Iris Van der Tuin, often inspired by Henri Bergson, Gilles Deleuze and Félix Guattari, Gilbert Simondon, Alfred North Whitehead) at times overlapping with the *speculative turn*, loosely gathered under the heading of speculative realism (largely its offshoot object-oriented ontology, led by Graham Harman and Timothy Morton, who themselves drew more from Martin Heidegger and Derrida, respectively). The nonhuman turn in general focused on "decentring the human in favour of a turn toward and concern for the nonhuman, understood variously in terms of animals, affectivity, bodies, organic and geophysical systems, materiality, or technologies" (Grusin vii). This momentary return to "theory," however, was always operating in the shadow of another "anti-theoretical" counterturn—"material" in a different sense—seeking to cement the all-too-human neoliberal identitarian markers of race, gender and sexuality (and only very seldom, class, for essential reasons) at the forefront of the humanities. Indeed, one can only use the terms "theory" and "anti-theory" ironically here; as Claire Colebrook notes, if an emphasis on the lived experiences of ineradicably politicized human identities positioned itself "against" theory in the 1980s, this position "is now exemplary of what counts *as theory*. . . . What was once anti-theory—a reaction against the detachment of texts from any supposition of humanity or meaning—is now so mainstream, that the same argument can be rehearsed and become central to a defence of theory 'after theory'" (Colebrook 35). In *The Promise of Multispecies Justice*, for example, Sophie Chao et al. note that, in simultaneously addressing issues of social justice and interspecies care by way of intersectional politics, "many writers have since abandoned the 'nonhuman' altogether, since it implies the lack of something—like 'nonwhite.' As we conceptualize multispecies futures, it is important to keep the intergenerational legacies of colonialism and racism in mind" (Chao et al. 2). Seen from another angle, however, such gestures risk betraying a certain *Hatred of the Nonhuman* tacitly at work in the environmental humanities, much like Tim Dean and Oliver Davis diagnose a *Hatred of Sex* the heart of queer studies (Davis and Dean vii). Davis and Dean themselves build on Jacques Rancière's notion of a constitutive *Hatred of Democracy*: a hatred for the rabble, for the elite, but either way a hatred for the *wrong sort* of voter. In our case, a hatred for those nonhuman members of the parliament of the living that care nothing about their representation.

This political reaction-formation against the non- or posthuman has, in a deeply problematic sense, closed itself off from critical reflection on the human and its non-anthropocentric ethical responsibilities. As Braidotti remarks, "even radical critics of Humanism, with their emphasis on diversity and inclusion, do not necessarily or automatically tackle the deeply engrained habits of anthropocentric thinking" (Braidotti 10). Furthermore, this forecloses an adequate conceptualization of extinction, of the world without us. For Colebrook, "there was a time, and there will be a time, without humans: this provides us with a challenge both to think the world as it is *for us*, and yet remain mindful that the imagining of the inhuman world always proceeds from a positive human failure" (Colebrook 32–33). Because deconstruction *both* dismantles the metaphysics of anthropocentric thinking *and* re-articulates the challenge of thinking the world *for us* with that of imagining it without us, hinging these over an impossible, differential abyss of convergence and compartmentalization, its ecological implications impose themselves all the more insistently today. In this sense, the *lack* intimated in the syntagm "nonhuman" may disclose another, nonconceptual line of inquiry all the more necessary at this very moment.

Eco-deconstruction, if such a thing exists, might be said to take place as the hiatus *between* environmental politics, the environmental humanism of other humans, *and* a more radical sense of posthumanist ecological ethics, of ecology as an ethics of the nonhuman. The earliest paper, to my knowledge, addressing what came to call itself "eco-deconstruction" grew out of such concerns. Robert Briggs's 2001 article "Wild Thoughts: A Deconstructive Environmental Ethics?," built on previous missed encounters between postmodernism and environmental ethics to argue that if deconstruction stipulates that any ethico-political decision takes place within a *context*, we have good reason to think of this context as *environmental*, both on the basis of Derrida's own work and of other scholarship in environmental ethics. Deconstruction, Briggs writes, "is worth considering, responding to, and taking seriously, not simply for its putatively 'critical' function, but also insofar as deconstruction may—*before any other consideration*—affirm the work of environmental ethics, and may affirm, therefore, the *necessity* of offering such work consideration" (Briggs 117). David Wood's 2007 paper "Spectres of Derrida: On the Way to Econstruction" likewise begins by articulating the tension between (individual) animal rights and broader environmental issues onto a wider field. Deconstruction, he argues, allows us to recognize "our original dependency on the whole life process, both in an evolutionary sense, and in terms of our current sustenance[,] an important counterweight to the humanism implicit both in a rigid economy of exchange, and in understanding the space of the ethical in terms of a pure gift" (Wood 265–56). The question of humanism, he adds, "arguably taints even the very word 'environment,' in its suggestion that we concern ourselves with what surrounds *us*" (Wood 266). An important swathe of eco-deconstruction grew out of this more philosophical impulse, culminating in the 2018 collection *Eco-Deconstruction: Derrida and Environmental Philosophy*.

Concurrently, often in tandem with this philosophical strain, bloomed a more ecocritical, "theory"-based interpretation of deconstruction, most visible in the special issues of the *Oxford Literary Review* edited by Timothy Clark: *Deconstruction, Environmentalism and Climate Change* (2010), *Deconstruction in the Anthropocene* (2012), *Overpopulation* (2016), and *What Might Eco-Deconstruction Be?* (2023). One should again note the influential work of Timothy Morton, whose *Ecology without Nature* was "inspired by the way in which deconstruction searches out, with ruthless and brilliant intensity, points of contradiction and deep hesitation in systems of meaning. If ecological criticism had a more open and honest engagement with deconstruction, it would find a friend rather than an enemy" (Morton, *Ecology Without Nature* 6). Morton's more recent attempts to integrate eco-deconstructive insights with those of object-oriented ontology in essays such as "Ecology without the Present" and "Ecology as Text, Text as Ecology" also merits mentioning, as does what Clark calls the "inhumanism" of Tom Cohen, Colebrook and J. Hillis Miller, developed in conversation with—but also deeply critically of—ecological interpretations of deconstruction. Nonetheless, these convergences and divergences, even outright disagreements, broadly point to the fruitfulness of eco-deconstruction. As Matthias Fritsch, David Wood and I wrote in the introduction to *Eco-Deconstruction*,

> if the prefix *eco-* points to the significance of natural/environmental context, it might be said that the distinctive thrust of eco-deconstruction is to affirm the significance of context, both "on the ground," as it were, and methodologically. This is the most plausible reading of Derrida's oft-quoted early claim that "*il n'y a pas de hors-texte* [there is nothing outside the text (*there is no outside-text*)]. But it needs to be added that no such context is ever fully saturated—that is to say, specifiable or determinable. There is no "final analysis," no ultimate frame. (Fritsch et al. 5)

Let us say that in sum, however, eco-deconstruction, like the environmental posthumanities, engages an undecidable hiatus between the politics of human environmental justice and a wider, ethical responsibility to nonhumans and their environments. Out of this abyss between the political and the ethical, eco-deconstruction invites us to ask if compartmentalization, disconnection, divergence, non-relationality, separation and withdrawal might not themselves lie at the core of a more radical ecological thinking. Consider the following citation from Derrida's "Avowing—The Impossible":

> The adverb "together" in the expression "living together" does not refer to the totality of a natural, biological or genetic ensemble, to the cohesiveness of an organism or of some social body (family, ethnic group, nation) that would be measured with this organic metaphor. "Living together" supposes, therefore, an

> interrupting excess both with regard to statutory convention, to law and with regard to symbiosis, to a symbiotic, gregarious or fusional living together. I would go so far as to say . . . that all "living together" that would limit itself to the symbiotic or that would be regulated according to a figure of the symbiotic or the organic is a first lapse of the sense and of the "must" of "living together." . . . "Living together" is reducible neither to organic symbiosis nor to the juridico-political contract. Neither to "life" according to nature or birth, blood or soil, nor to life according to convention, contract or institution. ("Avowing" 26–27)

How might one rethink the ethics of hospitality according to this puzzling figure of "living together"? Derrida's *Hospitality* seminars argue that the relation to the other that constitutes the ethical figure of hospitality must be founded upon an absolute, infinite interruption and non-relation, thus heralding "a pluralism of radical separation, a pluralism within which plurality is not that of a total community or of a cohesion or coherence of the whole, the 'coherence of elements constituting plurality.' It is plurality as peace" (*Hospitalité II* 74). The plurality of coherent elements is merely the plurality of the political, of politics, of war, which for Levinas is always secondary. The hospitality Levinas imagines instead is founded upon a peace that does not belong to the order of the political. This peace, however, is neither "natural" nor institutional. This rejection of "nature" owes itself to Levinas's peculiar understanding of the notion, where it is always depicted in terms of rootedness, of the pagan sacrality of the site. Nature, the political, rootedness, the site—all of these belong to the movement of *gathering* and *recollection* that underlies the Greek metaphysics of being. This gesture of gathering also constitutes the ontological figure of dwelling, which Derrida and Levinas invite us instead to think on the basis of "anachoresis," errancy, exile, expropriation, rootlessness, wandering. Every oikological movement of gathering, for deconstruction, presupposes an anterior movement of absolute withdrawal as its very condition. Against this Greek tradition of homecoming and repatriation, Levinas draws his metaphysics from the Bible, citing Leviticus 25:23; "The land shall not be sold for ever: for the land *is* mind; for ye *are* strangers and sojourners with me." One cannot possess any *oikos*, one is only ever an uninvited guest or a visitor therein. It is thus necessary to think ecology as "a chosen home and not a natural site. And since it is 'chosen.' . ., it is chosen on the basis of an erring. A root is not chosen. The root is a nature that is not chosen. A chosen home is the experience of an erring" (*Hospitalité II* 72n3). Levinas's ethics enable us to think an ecology that belongs to *no one*, foreclosing all politics of identarian belonging; "because there is no property, because there is only exile, expropriation, foreignness, because of that, this foreignness brings one closer rather than distancing" (*Hospitalité II* 300).

This is the sense of living together on the basis of disjuncture and dislocation noted above, one that reaches the limits of the human itself. As Levinas puts it in *Humanism of the Other*, "non-essence of the human, possibly less than nothing. 'It may be,' writes Blanchot, as one likes to declare, that 'the human passes.' It passes, it has already passed, to the extent that he was always appropriated to his own disappearance" (Levinas 67). This passing of the human, I believe, is how we ought to understand eco-deconstruction as post-humanism of the other. "Levinas wants humanism, but a humanism that does not refer to an essence, to a self-identity of the human, a humanism of what he calls 'the other human [*l'autre homme*],' of a human who is only a human where it has no site, where it is not at home and where it has no property or essence" (*Hospitalité II* 301).

I stated above that eco-deconstruction designates the logic by which any eco-concept must open itself up to *an* other that is no longer *its* other. Derrida indeed refers to the very stakes of any deconstruction as follows: "the difference between something like 'its' other and another that is no longer even reappropriable as 'its' other . . . , difference, then, between hospitality offered to *one's* other . . . and hospitality offered to an other that is no longer, has never been the 'its other' of dialectics" (*Hospitalité II* 149). But this implies the necessity of going beyond even a post-humanism of the other human. Any environmental humanities that would limit itself to the politics of human dwelling would constitute a first lapse in the sense of living together with others on earth.

REFERENCES

Braidotti, Rosi. *Posthuman Knowledge*. Polity, 2017.

Briggs, Robert. "Wild Thoughts: A Deconstructive Environmental Ethics?" *Environmental Ethics* vol. 23, Spring 2001, pp. 115–134.

Bryant, Levi et al. *The Speculative Turn: Continental Materialism and Realism*. Re.press, 2011.

Chao, Sophie et al. *The Promise of Multispecies Justice*. Duke University Press, 2022.

Clark, Timothy, ed. *Deconstruction in the Anthropocene*, special issue of *Oxford Literary Review* vol. 34, no. 2, December 2012.

———. *Deconstruction, Environmentalism and Climate Change*, special issue of *Oxford Literary Review* vol. 32, no. 1, July 2010.

———. *Overpopulation*, special issue of *Oxford Literary Review*, vol. 38, no. 1, July 2016.

Clark, Timothy and Philippe Lynes, eds. *What Might Eco-Deconstruction Be?*, special issue of *Oxford Literary Review*, vol. 45, no. 1, July 2023.

Cohen, Tom et al. *Theory and the Disappearing Future: On de Man, On Benjamin*. Routledge, 2012.

———. *Twilight of the Anthropocene Idols*. Open Humanities Press, 2016.

Colebrook, Claire. *Death of the PostHuman: Essays on Extinction, Vol. 1*. Open Humanities Press, 2014.

Daigle, Christine. "Environmental Posthumanities" in Stefan Herbrechter et al., eds., *Palgrave Handbook of Critical Posthumanism*. Palgrave Macmillan, 2022.

Davis, Oliver and Tim Dean. *Hatred of Sex*. University of Nebraska Press.

Derrida, Jacques. "Avowing—The Impossible" in Elisabeth Weber, ed., *Living Together: Jacques Derrida's Communities of Violence and Peace*. Fordham University Press, 2013.

———. "Deconstruction and the Other" in Richard Kearney, *Debates in Continental Philosophy: Conversations with Contemporary Thinkers*. Fordham University Press, 2004.

———. *Hospitalité, Volume II: Séminaire (1996–1997)*. Seuil, 2022.

———. *Margins of Philosophy*. Trans. Alan Bass. University of Chicago Press, 1982.

———. *Of Grammatology*. Trans. Gayatri Chakravorty Spivak. Johns Hopkins University Press, 1974.

———. *Points . . . Interviews, 1974–1994*. Trans. Peggy Kamuf et al. Stanford University Press.

———. *Psyche: Inventions of the Other, Volume II*. Eds. Peggy Kamuf and Elizabeth Rottenberg. Stanford University Press, 2008.

Fritsch, Matthias et al. *Eco-Deconstruction: Derrida and Environmental Philosophy*. New York: Fordham University Press, 2018.

Grusin, Richard, ed. *The Nonhuman Turn*. University of Minnesota Press, 2015.

Levinas, Emmanuel. *Humanism of the Other*. Trans. Nidra Poller. University of Illinois Press, 2006.

Lynes, Philippe. *Futures of Life Death on Earth: Derrida's General Ecology*. Rowman & Littlefield International, 2018.

Morton, Timothy. "Ecology as Text, Text as Ecology." *Oxford Literary Review* vol. 32, no. 1, 2010, pp. 1–17.

———. *Ecology Without Nature: Rethinking Environmental Aesthetics*. Harvard University Press, 2007.

———. "Ecology Without the Present." *Oxford Literary Review*, vol. 34, no. 2, 2012, pp. 229–238.

Neimanis, Astrida, et al. "Four Problems, Four Directions for Environmental Humanities: Toward Critical Posthumanities for the Anthropocene." *Ethics and the Environment*, vol. 20, no. 1, Spring 2015, pp. 67–97.

Van Dooren, Thomas, et al. "Multispecies Studies: Cultivating Arts of Attentiveness." *Environmental Humanities*, vol. 8, no. 1, pp. 1–23.

Wood, David. "Spectres of Derrida: On the Way to Econstruction" in Laurel Kearns and Catherine Keller, eds. *Ecospirit: Religions and Philosophies for the Earth*. Fordham University Press, pp. 264–287.

Chapter 9

Environmental Justice

The Beginning, Present and What Awaits Humanity in the Future

Erden El

Environmental justice, which began as a social movement, is one of the most significant eco-concepts of our time. It argues that environmental disasters affect impoverished communities, especially minority groups, disproportionately. This movement, which started as a reaction to the establishment of factories and facilities that pollute nature, particularly in the areas where economically disadvantaged citizens reside, defends the idea that environmental health is a human right. The basis of environmental justice is the rejection of the idea of privileged and unprivileged individuals. Based on the understanding that the environmental rights of the people who live in a rich neighborhood are the same as those of the impoverished living in a ghetto, this movement opposes social and transnational hierarchies. Environmental justice is an eco-concept that refers to the environmental rights of individuals in the places they dwell, work, and pray. Given that it is often the disadvantaged minorities living in ghettos who are deprived of environmental justice, the eradication of injustice becomes crucial. Although it initially set out to defend the rights of poor minorities living in ghettos, environmental justice is now understood as the natural right of every individual to live in a healthy environment. When people's environmental rights are at stake, it is crucial to transcend an ethical discourse and to make sure that these rights are guaranteed by legal regulations. For this purpose, this chapter will initially illustrate how the concept of environmental justice emerged, then it will touch on examples of the present, and finally, it will make a future projection of environmental justice. This chapter will argue that the term environmental justice should go beyond philosophy and authorities should put the eco-concept into their calculations and practices. The aim of the application of environmental justice must be to acknowledge that it is integrated not only in America and minority groups but also throughout the globe. It will provide advice on the process of transcending the philosophical discourse of the term and transforming it into legal regulations.

THE DEFINITION OF ENVIRONMENTAL JUSTICE

The definition of the term environmental justice has been made by the U.S. Environmental Protection Agency's (EPA) Office of Environmental Justice. According to EPA, environmental justice stands for "the fair treatment and meaningful involvement of all people regardless of race, color, culture, national origin, income, and educational levels with respect to the development, implementation, and enforcement of protective environmental laws, regulations, and policies" (EPA). EPA highlights ethnic origin and income while defining the term environmental justice. The reference to these concepts is meaningful since discrimination mainly takes place because of the aforementioned issues. Therefore, establishing environmental justice should stand for eliminating these discriminations based on income and color. The definition of EPA highlights the significance of equal rights and emphasizes the rights of disadvantaged individuals. As it can be understood from this definition, the emergence of the concept stemmed from the discrimination of these groups. The agency specifically focuses on fair treatment and defines that concept as well. The Office of Environmental Justice specifically opposes the negative environmental impacts affecting poor minorities and states that "no group of people, including a racial, ethnic or a socioeconomic group, should bear a disproportionate share of the negative environmental consequences" (EPA). So, it would be correct to state that environmental injustice mostly affects poor minorities. The concept of environmental justice emerged as a reaction to this environmental injustice.

According to Filamina C. Steady, "environmental justice challenges discrimination and disparities in the allocation of the benefits and burdens of economic development" (1). As can be understood from this definition of environmental justice, there is a reaction to uneven exposure to the dire outcomes of economic development between financially privileged and underprivileged groups. The poor are not asked for their opinions when establishing a health-threatening facility around them, but when it comes to seeing these damages, they are negatively affected by it. While they do not benefit from the development, they suffer from the harmful consequences. In response to this disproportionate situation, the environmental justice movement has emerged.

One of the most commonly accepted definitions of environmental justice has been made by the Asian Pacific Environmental Network (APEN) as "the right to a decent, safe quality of life for people of all races, incomes, and cultures in the environments where we live, work, play, learn and pray" (Environmental justice & API issues—what is EJ) and it defends the right of people to access a healthy environment wherever they are. At this point, the most important point is to remember that this is one of the most basic human rights. Robert R. Kuehn suggests four elements of environmental justice: "(1) distributive justice; (2) procedural justice; (3) corrective justice; and (4) social justice" should be applied to overcome environmental injustices (10681). According to Kuehn, distributive justice is the "equal . . . distribution of goods and opportunities" (10683). Procedural justice is "the right to treatment as an equal" (10688) and corrective justice is "fairness in the way punishments for lawbreaking are assigned" (10693). The adoption of these three ideals enables the establishment of social justice. In the establishment of environmental justice, these principles as well as the aforementioned principles should be adhered to. Any facility or company that violates the environment and public health should be punished equally, regardless of the income level and ethnicity of the injured party.

In the concept of environmental justice, "the environment and social justice" are intertwined (Murphy-Greene 2). If social justice is provided, environmental justice will be achieved. Michael Regan, an African-American who was elected to the presidency of the EPA in 2021, shared the following statements about the implementation of environmental justice: "We will be driven by our convictions that every person in our great country has the right to clean air, clean water, and a healthier life, no matter how much money they have in their pockets, the color of their skin or the community that they live in" ("EPA Head: 'Covid-19 Created a Perfect Storm for Environmental Justice Communities'"). As it is understood from Budryk's statements, it would be appropriate to think of environmental justice in terms of basic human rights.

THE EMERGENCE OF ENVIRONMENTAL JUSTICE

The emergence of the environmental justice movement can be attributed to the "African American community in Warren County, North Carolina" ("History of Environmental Justice"). However, it is impossible to pinpoint the exact date of the beginning. The movement in North Carolina emerged when the government chose the community as a landfill area ("History of environmental justice") which was very close to the community and would cause health problems among the residents. The main objection was that the landfill area was particularly selected since it was populated by an impoverished black community. Thus, it would be correct to state that the philosophy of the movement was a reaction to the toxification of the poor lands of communities- mostly if not wholly of color- with hazardous wastes. It was commonly believed that governments did not pay the necessary attention to the health of communities if the society in question were people of color.

Non-governmental organizations, minority groups, and religious institutions played a significant role in the movement. United Church of Christ Commission for Racial Justice was one of these groups which played an active role in the struggle against the injustices faced by black people in the region. As a commission, they started two projects examining whether or not the choice of the land as a landfill was related to the ethnic origin of the members of the community (Payne and Newman 259). These projects were important as they were the first projects in the United States to examine the possibility of a relationship between environmental pollution exposure and ethnicity (Payne and Newman 259). The results of the projects were striking since they described the possibility as strong. Although this did not lead to a change in the landfill location, it allowed the communities of color to have their voices heard in the following process. Although there were various protests before this movement in the early 1980s, it was insufficient to create public opinion. The movement of Christ Church, on the other hand, made a sound because it was organized and scientific, and encouraged the ensuing movements. As an outcome of the research and struggles, the concepts of environmental justice and environmental racism were introduced. The movement would later be called Environmental Justice (EJ). The term "environmental racism" was coined by Dr. Benjamin F. Chavis, "then-executive director of the United Church of Christ Commission for Racial Justice" (Jeffreys 678). According to Chavis, environmental racism stands for:

> Racial discrimination in environmental policy-making and enforcement of regulations and laws, the deliberate targeting of communities of color for toxic waste facilities, (and) the official sanctioning of the presence of life-threatening poisons and pollutants for communities of color. (Chavis vii)

According to Chavis' definition, the discrimination of ethnic groups in environmental policies is not accidental but deliberate. Because of this intention, it is associated with racism. While choosing landfill and similar places (where toxic wastes are placed), selecting places where poor people of color live is defined as a deliberate act by Chavis. Considering that the aforementioned wastes are defined as "life-threatening," it is possible to reach the interpretation that black communities are directly targeted by the authority. The avoidance of actions that can be positive as well as the implementation of negative actions can be considered racism within the process of enacting environmental laws concerning poor ethnic communities. In this case, the core question should be: *would the same environmental waste be dumped here again if the area in question was a place where wealthy white communities lived?*

Although income is a significant determinant, it is also stated by scholars that "statistics show that race is a better indicator than income in determining the probability that a community is polluted" (Collins 41). It has also been stated in the document *Toxic Wastes and Race in the United States* that "race was consistently a more prominent factor in the location of commercial hazardous waste facilities than any other factor examined" (cited in Collins 41). It is more probable that race plays a more significant factor in environmental justice. However, regardless of the discussion on whether race or income plays a more crucial role in this process, the main point to focus on is that there is undeniable inequality, and it should be eliminated.

AN EXAMPLE OF ENVIRONMENTAL (IN)JUSTICE: ALTGELD GARDENS CHICAGO

As Adamson, Evans, and Stein remark, "Environmental justice movements call attention to the ways disparate distribution of wealth and power often leads to . . . the unequal distribution of environmental degradation" (5). Environmental justice movements encourage research into the social and economic underpinnings of the disproportionate distribution of pollution. In his article "The Threat of Environmental Racism," Robert D. Bullard questions whether race is a significant factor in environmentally related health issues and states that lead poisoning is frequently seen in African-American children living near waste landfills in the United States (24). According to Bullard, the basis of this environmental discrimination is race (24). Lead poisoning is essentially a preventable disease, and without exposure, the disease will not occur. But are the authorities simply insensitive, or is it deliberate? This question should be at the center of environmental justice research. It is clear that color is one of the bases of environmental discrimination, and one of the examples of environmental discrimination took place in Altgeld Gardens, Chicago. Landfills are "facilities for the disposal of solid waste" (EPA) and according to Bullard, "Chicago is . . . one of the most racially segregated big cities" (25). Altgeld Gardens has posed an environmental threat since it was built on a former landfill (Reyna 4). It is also known as a toxic doughnut because it is in the middle of an environment surrounded by toxic fields (Reyna 1). US involvement in the Second World War caused the need for the accommodation of war workers in areas like Chicago. It was originally intended to be a temporary facility by the Federal Housing Authority (FHA) and the Chicago Housing Authority (CHA), but later became permanent. Altgeld Gardens was built on this area that used to be a "waste dump" (Altgeld Gardens). Former landfills and the surrounding industrial areas threaten the lives of its residents. In the 1970s, many residents who lived here suffered from asthma and cancer (Altgeld Gardens). It can thus be inferred that race is a significant determinant in terms of environmental injustice in the region.

HOW TO OVERCOME ENVIRONMENTAL INJUSTICE

The best recommendation for the route map to overcome environmental injustice came from Robert D. Bullard. What makes Bullard's principles optimal is that they set out with the principle of both prevention and remediation. Naturally, although remediation is a necessary procedure, environmental problems must have occurred before remediation can take place. Taking precautions is even more valuable as it is an attitude that prevents the problem from occurring. However, finding a solution to a problem that has already occurred is just as valuable. The establishment of environmental justice, then, will bring with it social equity. Bullard proposes five principles against the distribution of toxic wastes, the exposure of impoverished people to environmental injustice and discrimination, the fact that color is a determinant in this, and its negative effects on public and environmental health:

1. guaranteeing the right to environmental protection;
2. preventing harm before it occurs;
3. shifting the burden of proof of contamination to polluters, not the residents;
4. obviating proof of intent to discriminate; and
5. redressing existing inequities (Bullard 15)

Rob Nixon defines "slow violence" as a type of violence that develops slowly and is therefore difficult to understand (2). Unlike the rapidly advancing and instantaneous types of violence, since this type of violence progresses insidiously, it is difficult to identify the perpetrators of this violence. It would be useful to evaluate the concept of environmental justice together with slow violence. If the concept of slow violence is somehow recognized by law, then those who perpetrate this type of violence will not be able to carry out this behavior comfortably; they will not take it for granted that they will go unpunished. The lack of legal regulations, which facilitates the infliction of slow violence, simply encourages its practitioners to maintain this invisible brutality. Therefore, this research firstly recommends that slow violence should be recognized and punished by legal means. In this way, it will be possible to apply Bullard's first principle.

In the discussions of environmental justice, distributive justice is a key topic. Because it is the financially strong states and companies that produce the waste and it is the impoverished (often referred to as minority) communities that are exposed to the waste, distributive justice is directly related to environmental justice. Since distributive justice is "the equitable distribution of environmental risks and benefits" (Schlosberg 79) risks and benefits should be equally distributed. However, risks are distributed to disempowered communities, while benefits are shared by the wealthy in the existing state of affairs.

In terms of slow violence and distributive justice, it is obvious that legal measures should be taken in favor of the public. Once the legal basis has been established, it will be possible to realize Bullard's first principle. The people exposed to disproportionate toxic waste in the Global South, disadvantaged people living in ghettos, and all citizens of the world facing environmental injustice should first be made aware that environmental justice is a right. Environmental protection must be guaranteed by the law.

Although it sounds utopian, it is obvious that the most effective way to avoid harm is to prevent it before it occurs. In case a facility is built in a region that will generate toxic waste, local residents should be consulted. It should be taken into consideration that this facility will pollute the environment, causing nature to be adversely affected. As a result, people will experience health issues, and a lot of medical expenses will be incurred. The facility will most probably be closed in the process following the lawsuit filed by the public. However, if this facility is never opened from the beginning, both health and material damages will be prevented before they occur.

As Bullard's third principle indicates, the burden of proof belongs to companies, not residents. It is very important to realize this shift because it is not right to engage the people who are already suffering from environmental problems with legal processes. In this case, if there are complaints from the people in the environment, the authorities should immediately contact the companies and ask them to prove that they have not caused any damage to the environment. In this case, local administrations should appoint an inspector and if it is determined that the facility pollutes the environment at a dangerous level, it is necessary to impose sanctions with penalties up to the closure of the facility. If companies are cognizant of their financial liability for the treatment of people suffering from environmental diseases and rehabilitation of the environment, they will have to put public health at the forefront.

Bullard's fourth and fifth principles are interrelated, and the only effective way to eliminate existing injustices is to apply the fifth principle. Although it is known that the ideal is to prevent the emergence of the problem, it is certainly equally important to eliminate the problems that have already arisen. The working principle of environmental justice is to ensure that people do not suffer from environmental diseases anywhere in the world and to ensure that they live in a sterile environment. At this point, the existing problems should be eliminated so that everyone has equal conditions regardless of race, income, and ethnic origin.

CONCLUSION AND SUGGESTIONS

This study has discussed whether income and race play a role in environmental justice and has underlined the idea that environmental justice is related to income and ethnic origin and obviously certain reformative steps need to be taken. Foremost among these, environmental justice, which is identical to basic human rights, needs to go beyond the philosophical dimension and reach the legal dimension. There are various agencies, primarily the EPA, established for environmental justice. However, it would be meaningful to increase their number and impact. People who want to defend their environmental justice rights, not only in the United States but anywhere in the world should be able to apply to these institutions.

As stated before in this study, the most cost-effective solution is to prevent harm before it happens. If the damage can be prevented before it occurs, huge costs that will occur in the future can be avoided. Former landfills are not recommended to be used as residential areas. In case of necessity, it can be considered that an effective cleaning should be applied so that the area can be used. If it is proven with evidence that environmental injustice is caused by environmental racism, then more legal penalties are required. Thus, companies and institutions that may attempt to do so will refrain from it if they consider the probable costs.

Bullard's principles related to environmental justice can be a good route map for lawmakers. As with Bullard's first principle, viewing environmental protection as a right and guaranteeing it by law may be the most effective step in the solution. The application of this principle can work in two ways. First, it ensures the protection of the person who may be exposed to environmental injustice; second, it may cause institutions and organizations to avoid causing harm. As for the second principle, it is cost-effective as mentioned earlier. Being cost-effective should not mean that large budgets cannot be allocated to places where environmental injustice has occurred. If necessary, it is possible to establish superfunds. Superfund areas became popular in the 1980s when the public reacted to contaminated areas and, "in response, Congress established the Comprehensive Environmental Response, Compensation and Liability Act (CERCLA) in 1980" ("What Is Superfund?"). CERCLA is referred to as a superfund among the public and stands for cleaning a contaminated area by investing a grand budget ("What Is Superfund?"). The establishment of superfund areas is a very effective method for eliminating the negative effects of environmental pollution but requires a large budget. It is worth remembering here that if the damage is prevented without the need for superfunds to be established, it is more advantageous for all parties, both in America and elsewhere in the world. If this can be achieved, people, animals, and nature will not be harmed, and great costs will be avoided. However, if superfunds need to be installed, it's also essential to make it happen. The application of the third principle will also prevent people dealing with environmental problems from taking extra responsibility. People who deal with environmental problems and health issues should not have to prove that there is contamination in the area. Otherwise, they will either become fatigued or refrain from seeking their rights because of this mobbing, or they will give up. So, as in the Bullard principles, the obligation to prove contamination should belong to the companies. As stated in the interconnected fourth and fifth principles, the possibility of discrimination should be eliminated, and equality should be established.

Future expectations regarding environmental justice are the recognition of environmental justice as a rule, its evaluation within the scope of fundamental rights and freedoms, and its evolution from an ethical dimension to a legal dimension. It is well-known how important education is in this regard. For this reason, the concept of environmental justice can be integrated into school curricula. The importance of environmental awareness should be transferred to posterity.

REFERENCES

Adamson, Joni, et al. "Introduction: Environmental Justice Politics, Poetics, and Pedagogy." *The Environmental Justice Reader: Politics, Poetics, & Pedagogy*, by Joni Adamson et al. University of Arizona Press, 2002.

Budryk, Zack. "EPA Head: 'Covid-19 Created a Perfect Storm for Environmental Justice Communities.'" *The Hill*, March 24, 2021, retrieved December 3, 2023,https://thehill.com/policy/energy-environment/544702-epa-head-pandemic-created-a-perfect-storm-for-communities-affected/#.YFyVCLiC-N4.twitter%20.

Bullard, Robert D. "Overcoming Racism in Environmental Decisionmaking." *Environment: Science and Policy for Sustainable Development*, vol. 36, no. 4, 1994, pp. 10–44. doi: 10.1080/00139157.1994.9929997.

———. "The Threat of Environmental Racism," *Natural Resources & Environment*, Winter 1993, vol. 7, no. 3, Facility Siting (Winter 1993), pp. 23–26, 55–56

Chavis, B.F.J. "Preface." *Unequal Protection: Environmental Justice and Communities of Color.* Ed. Robert Doyle Bullard. Sierra Club Books, 1994.

Collins, Robin. "Environmental Equity and the Need for Government Intervention: Two Prospects," *Environment*, vol. 35, no. 9, 1993, pp. 41–43. doi: 10.1080/00139157.1993.9929127.

"Environmental Justice & API Issues—What Is EJ." *APEN*, retrieved December 3, 2023, http://archive.apen4ej.org/issues_what.htm.

EPA. Basic Information about Landfills, retrieved December 3, 2023, https://www.epa.gov/landfills/basic-information-about-landfills.

———. Environmental Protection Agency, https://www.epa.gov/environmentaljustice/ej-2020-glossary.

"History of Environmental Justice." *History of Environmental Justice | the University of Michigan School for Environment and Sustainability*, retrieved December 3, 2023, https://seas.umich.edu/academics/master-science/environmental-justice/history-environmental-justice.

Jeffreys, Kent. "Environmental Racism: A Skeptic's View." *Journal of Civil Rights and Economic Development*, vol. 9, no. 2, Spring 1994, pp. 677–691.

Kuehn, Robert. "A Taxonomy of Environmental Justice Issues." *Environmental Law Reporter* vol. 30, 2000, pp. 10681–10703.

Murphy-Greene, Celeste. "Introduction." *Environmental Justice and Resiliency in an Age of Uncertainty.* Routledge, 2022.

Nixon, Rob. *Slow Violence and the Environmentalism of the Poor.* Harvard University Press, 2011.

Payne, Daniel, and Richard Newman. "United Church of Christ Commission for Racial Justice." D.G. Payne et al., eds., *The Palgrave Environmental Reade*r. Palgrave Macmillan, 2005.

Reyna, Lacy. "Fight or Flight: The Altgeld Gardens Conflict," SUST 240 Waste & Consumption Roosevelt University December 2014, retrieved December 27, 2022, https://futureofschaumburg.files.wordpress.com/2015/02/reyna-ej-final.pdf.

Schlosberg, David. "The Justice of Environmental Justice: Reconciling Equity, Recognition, and Participation in a Political Movement." *Moral and Political Reasoning in Environmental Practice*. Ed. A. Light and A De-Shalit. MIT Press, 2003.

Society of Architectural Historians. "Altgeld Gardens," retrieved December 3, 2023, https://sah-archipedia.org/buildings/IL-01-031-0026.

Steady, Filamna C. "Introduction." *Environmental Justice in the New Millennium Global Perspectives on Race, Ethnicity, and Human Rights*. Palgrave Macmillan, 2009.

"What Is Superfund?" Environmental Protection Agency, retrieved December 3, 2023, www.epa.gov/superfund/what-superfund.

Chapter 10

Ecopsychology and Indigenous Ecosophy

Lessons in Sustainability

Panchali Bhattacharya & Pritam Panda

Ecopsychology and indigenous ecosophy are two essential threads that intersect in the intricate fabric of our contemporary society and provide deep lessons in sustainability. These strands highlight the innate bond between the natural world and the human mind. Understanding this interaction is crucial as mankind navigates the problems of the twenty-first century, which are typified by the unrelenting march of technology and unsustainable development. In order to enhance the resonance of these intricately woven strands, this paper sets out to explore the rich tapestry that ecopsychology and indigenous ecosophy have woven together:

> Indigenous traditional ecological knowledge may offer a way of radically, but usefully, re-storying modernity in a manner that may be profoundly useful to those climate change and other environmental campaigners who are now seeking a post-secular environmental psychology that moves from facts to emotion, from the head to the heart. Given the urgency of our current environmental predicaments, such re-storying and reframing may be timely. (Coope 156)

The purpose of the present study is to unveil the transformative potential of uniting the wisdom of indigenous communities with the insights of ecopsychology. In doing so, we aim to shed light on how this synthesis can guide us toward a more sustainable coexistence with the planet that sustains us. Ecopsychology, an interdisciplinary field, serves as our compass in this exploration. It melds the realms of ecology, which concerns the delicate harmony of life on Earth, and psychology, which delves into the intricate workings of the human mind. In weaving together these domains, ecopsychology seeks to reposition the human psyche within the broader environmental framework.

We are on the verge of a major environmental catastrophe as the unrelenting progress of industrialization continues to deplete the resources of nature, prioritizing profit above everything else. The anxieties of dwindling resources, vanishing species, burgeoning populations, and a planet increasingly inhospitable to life weigh heavily upon our collective consciousness: "Worries of a lack of resources, of a decreasing number of species, of a booming human population, and of an overall place to live have been great concerns in the highly destructive civilization which we live" (Ambrosius 1). The emergence of ecopsychology is a response to the growing dissonance between humanity and the environment, a dissonance that has given rise to destructive behaviours with far-reaching consequences. As our separation from nature deepens, so does our peril. Ecological crises manifest not only in environmental degradation but also in the human psyche. In fact, global climate change has massive bearings on human psyche which often leads to significant behavioural changes in an individual, including anger management issues, anxiety disorder, or even depression. As an effect of ecological change, human beings tend to demonstrate ecologically destructive behaviour that wreaks havoc on the overall well-being of society. By separating ourselves from nature, we not only threaten the nourishment of nature, but our own welfare as well.

The discussion on the inextricably entwined relationship between these two dimensions was identified for the first time by Sigmund Freud, in *Civilization and Its Discontents* (1929). In this book, Freud has acknowledged the deep correlation that exists between human mind and the external environment. He states that "Our present ego-feeling is, therefore, only a shrunken residue of a much more inclusive—indeed, an all-embracing—feeling which corresponded to a more intimate bond between the ego and the world about it" (Freud 15). This nascent ecological concept started taking further impetus in the early 1960s with Robert Greenway's research on *psychoecology* which theorized that "the mind is nature, and nature, the mind" (Greenway 131). According to his opinion, ecopsychology can effectively function as a tool for better understanding the human-nature relationship. It can also help to identify the issues that are adversely affecting this man-nature dynamic and suggest practical ways that can accelerate the process of healing (Greenway 122). However, as a formalised domain of ecocritical theory, it was Theodore Roszak who brought the concept of ecopsychology into wide usage by emphasizing the various ways in which the Earth suffers due to its unrestrained exploitation by its inhabitants which, in turn, significantly disrupts the natural flow of human life as well (Roszak 3). Paul Shepard's view on ecopsychology, in his 1982 volume, *Nature and Madness*, is equally intriguing as it elucidates, with relevant contemporary instances, the impact that our gradually failing association with nature had upon our emotional as well as psychological development. Shepard has aptly pointed out that "Human sanity requires some less-than-obvious connections to nature as well as the necessities of food, water, energy, and air. We have hardly begun to discover what those connections may be" (Shepard 122).

In India, the cultural ethos of living in harmony with nature has been reflected in the ancient scriptures like the *Isopanishad*, the *Hitopadesha,* the *Panchatantra,* the *Shuka-saptati* or Kalidasa's *Meghadutam* and *Ritusamhara* (Dey 4873). Nature has been attributed with divine qualities in these texts, and, therefore, the alignment with nature is a spiritual journey rather than a purely behavioral one. If we consider the religious rituals exemplified in these scriptures, we will find that most of the rituals are intrinsically linked with the celebration of the five elements that constitute our human body. In these discourses, nature has been portrayed as a medium of mood upliftment. Revisiting the ancient mythological models that blend the ecological needs with the psychological needs is also significant in this context. The ancient *gurukul* system in India was built around a similar model where dissemination of knowledge was facilitated in proximity with nature. The *gurukul* system allowed the students (known as shisyas) to nurture their skills "in a natural surrounding where the shisyas lived with each other with brotherhood, humanity, love, and discipline. . . . All these helped in the personality development and increased their confidence, sense of discipline, intellect and mindfulness which is necessary even today to face the world that lay ahead" (Chandwani). In the modern times, Rabindranath Tagore reinvigorated the same spirit of education by developing and expanding *Shantiniketan*, a university town where the conventional classroom-based teaching-learning arrangement is debunked in favour of nature-oriented open-air education system. As psychological well-being is directly linked with the capacity to learn independently, these ancient educational prototypes bear immense significance for the development of future courses of education.

As Henry David Thoreau suggests, human beings experience perceptible and verifiable psychological benefits when they come in close contact with nature and gives voice to their inherent desire to engage with the world of flora and fauna: "Yet I experienced sometimes that the sweetest and tender, the most innocent and encouraging society may be found in any natural object, even for the poor misanthrope and most melancholy man. There can be no very black melancholy to him who lives in the midst of Nature and has his senses still" (Thoreau 101). This psychological relationship with the surrounding natural world is purely organic. There is a natural restorative power embedded in the environment which is in complete synchrony with the well-being of the human mind and body. That is why Hartig has pertinently noted that the devastating psychological effects of natural and manmade environmental disasters often lead to posttraumatic stress disorder (PTSD), anxiety, fear, anger, and even deep-rooted aversion towards the natural objects. This connection has undoubtedly intrigued human minds, largely jaded by the incongruence of human existence in the present times. Thus, the insights from environmental psychology are taken into consideration by therapeutic practitioners as well in the twenty-first century who explore the human-nature association in order to rightly understand the emotional state of human mind. According to Roszak, the concept of ecopsychology is akin to the idea of *biophilia hypothesis* which believes that there has always been an inevitable attachment between human beings and other forms of living organisms. In fact, human beings are at all times driven by a subconscious biological "urge to affiliate with other life forms" (Kellert and Wilson 416), and with nature as a whole. Ecopsychology, like the biophilia hypothesis theory, advocates a love for Nature that goes beyond the realm of romantic sentimentality. If we look at health in the holistic sense, biophilia hypothesis propagates the idea that propinquity with nature can boost our immune systems, support mental and emotional health, create social connections, aid physical recovery, and improve cognitive performance. Ecopsychology operates in close proximity with this same paradigm, and hence proves to be a key eco-concept that can ensure ecological preservation and sustainability.

IDENTIFYING ECOPSYCHOLOGY IN INDIGENOUS ENVIRONMENTAL PHILOSOPHY

It is quite evident that one of the most crucial challenges of the twenty-first century is to ascertain the stability of the human race against the overwhelming onslaught of modernization and unsustainable development. Environmental theorists tend to focus majorly on the search for "scientific, technological, and increasingly, economic solutions" (Kluwick 502). However, such an approach undoubtedly has its own limitations, since "scientists themselves are increasingly aware that science and politics are unlikely to be enough to prompt global behavioral action in response to climate change" (Porter 145). Therefore, to mobilize large numbers of people for ecological change, it is imperative that an alternative ecological philosophy is introduced in the mainstream environmental pedagogical discussion. In the face of overarching anthropomorphic activities and a strictly growth-oriented development paradigm which poses serious threat to ecology, the ecosophy of the indigenous communities offer one such environmental perception in which human beings live together peacefully with nature and maintain a balance in the entire ecosystem. In the indigenous tradition, the knowledge of the natural world is not merely limited to the understanding of flora and fauna. It encompasses the knowledge of the land as a whole which includes the physical landscape as well as the entire bio-network. Aldo Leopold's concept of *land ethic* basically centres around this idea where he endorses the awareness of taking care of the people, the land, and most importantly, ensuring a strong bond between both. As a theoretical framework, the concept of a community is expanded by land ethic to comprehend not just humans but also the soil, water, plants, and animals of the world. As a result, it creates a biotic community in which every living being in the ecosystem depends on and communicates with one another, because "when we see land as a community to which we belong, we may begin to use it with love and respect" (Leopold viii), and this mutual respect leads to emotional and psychological well-being.

The sense of reverence for the land that the members of the native communities demonstrate further propagates a sense of belonging and responding to the cosmos which have some correlation to Ursula Heise's understanding of *sense of place*. It refers to the particular experience of an individual in a particular place which gives the setting a special personality. In fact, it is the propinquity of the indigenous peoples with the world of flora and fauna that escorts them to take up "the challenge . . . to shift the core of cultural imagination from a sense of place to a less territorial and more systemic sense of planet" (Heise 56). Thus, they can effortlessly comprehend their exact place and thereby successfully apprehend their value in the universal ecosystem. This realization of their worth and the role that nature plays in determining it help them to advocate and celebrate, unequivocally, the harmonious cohabitation of human beings and other species within a natural ecosystem. The environmental philosopher Arne Naess and the other propagators of the theory of deep ecology, like Bill Devall, and George Sessions, endorse a similar all-inclusive perception which claims that anthropocentric thinking has alienated human from the natural environment and instigates them to exploit nature:

> Deep ecology is rooted in two fundamental principles. The first principle is self-realization, which affirms that each individual (human or otherwise) is a part of larger whole, or Self, which encompasses ultimately the planet earth and the entire cosmos. . . . The second fundamental principle of deep ecology . . . is biocentric equality, which upholds that all elements of the biosphere have an equal right to live and flourish. (Christopher 128)

Such alienation from ecological roots creates the premise for the intervention of an eco-concept like ecopsychology.

The members of the indigenous communities do not preserve natural resources just because it provides sustenance. Preservation of the environment and its constituent elements is a spiritual practice for the members of these native groups—a ritualistic way of life (Gupta and Guha 18). Noted indigenous historian, Joy Porter corroborates this perception in the context of Native Americans by opining that "a community's spiritual framework plays an important role in informing responses to environmental change, whether that framework is simply part of the social and historical matrix of a community or whether it directly and obviously informs decision making" (Porter xvii). For the members of the indigenous groups, "the idea that land and the spiritual life of specific communities are indissolubly linked" (Porter xviii) stands true. In that sense, the indigenous peoples are true envoy of "ecospirituality" that evokes the "manifestation of the spiritual interconnection between human beings and the environment . . . [and which] engages a relational view of person to planet, soul to soil, and the inner to outer landscape" (Lincoln 228).

The cultural ecologist and philosopher, David Abram's *The Spell of the Sensuous: Perception and Language in a More-Than-Human World* (1996), is one of the first books to look closely at the ecological philosophy of diverse indigenous groups and their oral cultures. Abrams explains the sensory experiences as animistic in nature because its cosmology affords a material field of perception that is inherently animate and self-organizing. According to Abram, the indigenous people, through the medium of their oral culture, establish a relationship of senses with their surroundings, much unlike the physical attachment inherent in an alphabet-based culture. He analyzed the deep impact of the advent of formal writing systems that has, to a large extent, diluted the relationship of the indigenous peoples with the natural world that surrounds them. As exemplified by Abram in his absorbing treatise, indigenous communities like the Australian Aborigine and the Apache of the American Southwest depend for their existence on the language of their landscape. In spite of the displacement and cultural shift engendered by the process of rapid industrialization and growing urbanization, these communities have maintained a constant responsiveness to their surrounding ecological construct on a local scale, thereby effectively addressing the pressing needs of the planet. David Abram's ethnographic research in Southeast Asia and the Himalayas incorporates his close association with shamans, sorceress and healers which offered him a deeper understanding of the indigenous relationship between ecology and human psyche. He identified that the forest which nurtures and nourishes the tribal people, also casts a silent yet positive effect on the mindset of these people and endows them with clarity and prudence and develops them into a self-contained entity.

The ecological philosophy of the indigenous communities propagates the idea of worshipping various natural features that hold intimate feelings for them, thereby inducing a sense of *geopiety*. In his essay, "Notes on Early

American Geopiety," American geographer, John Kirtland Wright (1966) coined the term *geopiety* to describe "pious emotion evoked by the wonder and the terror of the earth in all its diversity" (Wright 1). The term basically implies action and notions of sensitivity and reciprocation towards a place that is considered sacred. It illustrates "a person's special attachment and reverence for particular places or locations in the environment: places that have specific, highly personal meaning" (Knowles 9). This feeling of reverence for a shared space figures prominently in the ecosophy of various tribal communities and is unanimously practiced and advocated as well:

> Sacred groves (SGs) are patches of trees on forest land that are protected communally with religious zeal and connotations. These forest areas have been protected since ages by traditional societies and indigenous communities with their socio-cultural and religious practices. Sacred groves, as a rule, are treated piously. . . . SGs carry direct and everlasting pious status and assist in maintaining social fabric of the society. (Kandari et al. 1)

The indigenous groups often ascribe divine connection to these seemingly ordinary landscapes as they evoke unique meanings for them, thereby substantiating their sense of geopiety. This close and timeless interaction between space, place and the human world, the intimate exchanges between the biosphere and the anthrop definitely bestow the landscape of the tribes with a spirit of its own.

The value of indigenous ecological philosophy, therefore, must be given primacy in ecopsychology as it holds a legacy that offers holistic and common sensible solutions to nurturing an ecosystem that is fit for human habitation as well as for all other forms of living entities that can mutually thrive. The indigenous communities throughout the world are the most prominent propagators of this invaluable knowledge; however, their understanding of ecological preservation has always been viewed by the dominant cultural manifestations as a sort of subordinated knowledge and rejected as insufficient. Native eco-sensibility is typically positioned below the requisite level of cognition and at the bottom of the socio-political hierarchy in the prevailing frameworks. This is because local knowledge, in contrast to the rationalist worldview of scientific data, is judged to be prejudiced and unacceptable. Nevertheless, Thomas Berry (2000), one of the most eminent cultural historians, has appropriately summarised the fundamental difference between these two worldviews: "Indigenous people live in a universe, in a cosmological order, whereas we, the people of the industrial world, no longer live in a universe. We live in a political world, a nation, a business world, an economic order, a cultural tradition, a Disney dreamland" (Berry 14–15). There is no doubt that incorporation of indigenous ecosophy in our everyday life can significantly minimize the negative consequences of environmental exploitation and help us feel more connected to our surrounding emotionally. However, it is interesting to note that the psychological element in indigenous environmental worldview has not been proliferated or documented widely till date. Consequently, indigenous ecosophy has not yet received adequate primacy in the conventional eco-psychological discourse. Nevertheless, the rich possibility embedded in this unification cannot be ignored altogether.

CONCLUSION

During the outbreak of the COVID-19 pandemic in early 2020, there was an astronomical surge in mental health issues, engendered by the forceful captivity of human beings within the confine of their residences. As Fernandez, Quic, and Abreu point out:

> This new phenomenon also generates, first of all, health crises and later, in a more concealed and insidious way, economic, ecological, and social crises. The systemic nature of the COVID-19 crisis and its unprecedented cascading effects have reverberated in all sectors and levels. . . . Furthermore, there has been and continues to be a breakdown at different levels: emotional, relational, and territorial. (Fernandez et al. 2)

Such an unforeseen catastrophe brought to the fore the close inter-relationship that human psychology shares with the natural world that they are surrounded by. It emphasized the necessity of going back to nature for physical as well as psychological well-being. The forceful lockdown of the citizens within the four corners of their rooms actually validated the relevance of ecopsychology in the contemporary era by showing "how to deal with alienation, intense stress, death, deep societal change, nature regeneration and emotional healing" (Fernandez et al. 2). In such a context, Andy Fisher's interpretation of ecopsychology holds credence: "Ecopsychology can be an alternative or corrective to the bureaucratic instrumentalism and psychological illiteracy of the mainstream environmental movement. . . . Ecopsychology seeks to overcome the dualisms and alienations at the root of our psychological and ecological problems" (Fisher 171). Modern ecopsychologists hold that engaging in earth-nurturing activities like environmental conservation or awareness campaigns, meditating outdoors, going on wilderness retreats, celebrating the seasons or other natural phenomena, participating in nature-based festivals, and spending time with animals—all have therapeutic benefits on the geriatric population as well as on those who have psychological special needs.

The modern ecological movements throughout the world majorly act on the basis of our fear, which should not be the main purpose of our efforts in creating a green environment. As such, these movements bestow a restrictive texture on the capacities of environment as an all-encompassing entity. In doing so, we often fail to benefit from the holistic approach that celebrates the relationship between humans and the environment. The indigenous environmental philosophy promotes such an all-rounded perception of nature by moving out from being a temporary environmental movement to an eternal mode of establishing congruence between human and non-human. As an ecological concept which bears an immense significance in the twenty-first century, ecopsychology should take into account this approach to the human mind and argue that human beings are products of the predominant social and cultural models in our society. The present understanding of ecopsychology needs to adapt to the magnanimity of the environment, which is not merely a solution provider, rather an entity that is capable of self-sustenance as well as human nourishment in the coming times. The dynamics associated with eco-psychological studies assume a dimension where we contrast the shrinking natural resources and the accentuating psychological anomalies in human society. Often regarded as a therapeutic technique, eco-psychology fundamentally promotes a sense of sustainable development in human minds by bringing people spiritually closer to nature and making them feel responsible for the well-being of the environment that nourishes them. However, people often demonstrate negligence in their sustainable practices, the primary reason being their belief that individual efforts cannot bring significant changes to the overwhelming ecological crisis at hand (Gifford 291). A response to this mindset can be identified in the indigenous worldview which majorly believes in sustainable usage of natural resources (Suzman xii). It basically offers a lesson that if we want to subvert the ecological catastrophes that are eventually going to engulf humanity, we need to immediately return to the indigenous worldview and blend it with our contemporary lifestyle in order to ensure sustenance (Narvaez et al. 11). The present study draws attention to this possible incorporation of indigenous ecosophy in the study of an emerging eco-concept like ecopsychology which can meaningfully contribute to mental stability required for a healthy existence.

REFERENCES

Abram, David. *The Spell of the Sensuous: Perception and Language in a More-Than-human World.* Pantheon Books, 1996.

Ambrosius, Wendy. "Deep Ecology: A Debate on the Role of Humans in the Environment." *UW-L Journal of Undergraduate Research,* vol. 8, 2005, pp. 1–8. https://www.uwlax.edu/globalassets/officesservices/urc/juronline/pdf/2005/ambrosius.pdf

Berry, Thomas. *The Great Work: Our Way into the Future.* Broadway Books, 2000.

Chandwani, Nikhil. "The Importance of the Gurukul System and Why Indian Education Needs It." *The Times of India*, March 8, 2019, https://timesofindia.indiatimes.com/blogs/desires-of-a-modern-indian/the-importance-of-the-gurukul-system-and-why-indian-education-needs-it/

Christopher, Daniel L. Smith. "'Deep Ecology' and Radical Environmentalism." In *Battleground: Religion,* vol. 1, pp. 127–132. Greenwood Press, 2008.

Coope, Jonathan. "How Might Indigenous Traditional Ecological Knowledge (ITEK) Inform Ecopsychology?" *Ecopsychology*, vol. 11, no. 3, 2019, pp. 156–161. doi:10.1089/eco.2019.0005

Dey, Anup Kumar. "Ancient Indian Perspective on Environment." *Our Heritage,* vol. 68, no. 1, 2020, pp. 4872–4881. https://www.academia.edu/68857577/Ancient_Indian_Perspective_on_Environment

Fernandez, Manuela, et al. "Indigenous Wisdom and Ecopsychology for Empowering Emerging Latin American Change Makers: Moving from Covid Disruption to Better Social and Ecological Wellbeing." *Global Environments Network*, Global Diversity Foundation, 2021. http://globalenvironments.org/wp-content/uploads/2021/09/Indigenous-Wisdom-and-Ecopsychology_Final-Report.pdf

Fisher, Andy. "Ecopsychology at the Crossroads: Contesting the Nature of a Field." *Ecopsychology*, vol. 5, no. 3, 2013, pp. 167–176. https://doi.org/10.1089/eco.2013.0031

Freud, Sigmund. *Civilization and Its Discontents*. Penguin, 2002.

Gifford, Robert. "The Dragons of Inaction: Psychological Barriers That Limit Climate Change Mitigation and Adaptation." *American Psychologist,* vol. 66, no. 4, 2011, pp. 290–302. https://psycnet.apa.org/doi/10.1037/a0023566

Greenway, Robert. "The Wilderness Effect and Ecopsychology." *Ecopsychology: Restoring the Earth, Healing the Mind,* edited by Theodore Roszak, Mary E. Gomes and Allen D. Kanner. Sierra Club Books, 1995, pp. 122–135.

Gupta, Abhik, and Kamalesh Guha. "Tradition and Conservation in Northeastern India: An Ethical Analysis." *Eubios Journal of Asian and International Bioethics,* vol. 12, no. 1, 2002, pp. 15–18. https://www.eubios.info/EJ121/EJ121F.htm

Hartig, Terry. "Three Steps to Understanding Restorative Environments as Health Resources." *Open Space: People Space*, edited by C. W. Thompson and P. Travlou. Taylor & Francis, 2007, pp. 183–200.

Heise, Ursula K. *Sense of Place and Sense of Planet: The Environmental Imagination of the Global*. Oxford University Press, 2008.

Kandari, Laxman Singh, et al. "Conservation and Management of Sacred Groves, Myths and Beliefs of Tribal Communities: A Case Study from North-India." *Environmental Systems Research,* vol. 3, 2014, pp. 1–10. https://doi.org/10.1186/s40068-014-0016-8

Kellert, Stephen R., and Edward O. Wilson. *The Biophilia Hypothesis.* Island Press, 1993.

Kluwick, Ursula. "Talking about Climate Change: The Ecological Crisis and Narrative Form." *The Oxford Handbook of Ecocriticism,* edited by Greg Garrard. Oxford University Press, 2014, pp. 98–117.

Knowles, J. Gary. "Geopiety, the Concept of Sacred Place: Reflections on an Outdoor Education Experience." *Journal of Experiential Education,* vol. 15, no. 1, 1992, pp. 6–12. https://doi.org/10.1177/105382599201500101

Leopold, Aldo. *A Sand County Almanac and Sketches Here and There.* Oxford University Press, 1949.

Lincoln, Valerie. "Ecospirituality: A Pattern That Connects." *Journal of Holistic Nursing,* vol. 18, 2000, pp. 227–244. https://doi.org/10.1177/089801010001800305

Narvaez, Darcia et al. "People and Planet in Need of Sustainable Wisdom." *Indigenous Sustainable Wisdom: First-Nation Know-how for Global Flourishing,* edited by Darcia Narvaez et. al. Peter Lang Publishing Inc., 2019, pp. 1–24.

Porter, Joy. *Native American Environmentalism: Land, Spirit, and the Idea of Wilderness*. University of Nebraska Press, 2014.

Roszak, Theodore. "The Voice of the Earth: Discovering the Ecological Ego." *The Trumpeter*, vol. 9, no. 1, 1992. http://www.icaap.org/iuicode?6.9.1.8

Shepard, Paul. *The Tender Carnivore and the Sacred Game*. University of Georgia Press, 2011.

Suzman, James. *Affluence Without Abundance: The Disappearing World of the Bushmen*. Bloomsbury Publishing USA, 2017.

Thoreau, Henry David. *Walden*. Pan Macmillan, 2016.

Wright, John Kirtland. "Notes on Early American Geopiety." Human Nature in Geography, vol. 45, Harvard University Press, 1966, pp. 250–285. https://doi.org/10.4159/harvard.9780674434639.c15

Chapter 11

Hyperobjects

How to Move Forward While Entangled in the Mesh

Ana Simić

What do black holes, capitalism, global warming, cities, and Styrofoam cups have in common? The answer is that they are *hyperobjects*, entities so vast both temporally and geographically that they are very hard or impossible to grasp in their entirety. We can perceive only slices of hyperobjects or their effects on us. The eco-concept of hyperobjects was first introduced by Timothy Morton, Professor of English at the University of California, a scholar who has written extensively on philosophy and ecology. This eco-concept has had a significant impact on philosophical and ecological discourse and become one of the trending topics in ecological thought and ecocritical theory. This chapter will critically examine the origin, meaning, development, current condition, and future projections of hyperobjects, as well as potential implications for ecological thought and approach to ecological challenges.

The term "hyperobjects" was coined by Timothy Morton and first appeared in his book *The Ecological Thought* (2010). Morton used the term to describe entities that possess characteristics that transcend ordinary perceptual and cognitive capacities. Throughout his book, Morton refers to phenomena such as global warming, radiation, or the Big Bang as examples of hyperobjects. Due to their temporal and geographical scale, they are not directly observable, only measurable by scientific instruments. Global warming cannot be directly observed, but its aspects, or manifestations, are experienced by humans and the biosphere as extreme weather events or sunburn on our skin. Therefore, hyperobjects can be perceived via their interactions with other entities. This interconnectedness is one of the major characteristics of hyperobjects, the other three being non-locality, viscosity, and temporal undulation.

The concept of hyperobjects was developed within the broader context of object-oriented ontology and speculative realism. Morton draws inspiration from philosophers like Graham Harman and Quentin Meillassoux, who challenge anthropocentric perspectives and emphasize the independent existence of objects. Speculative realism in philosophy rejects correlationism, or the Kantian subject-object relations, and promotes flat ontologies that do not extol the human mind over other entities (Campbell 1316). Object-oriented ontology claims that all objects, whether physical or abstract, have an independent and irreducible existence, in contradiction to some traditional philosophical approaches that often privilege human perception and human-centric perspectives. Objects possess their own intrinsic qualities and are not merely reducible to their relationships with humans or other objects. Therefore, human, non-human, and abstract objects are treated as coequal. This philosophical school thus rejects most ethical systems' correlationist and anthropocentric tendencies (Bricker 359). In object-oriented ontology, the focus is on the objects themselves, rather than on how they are experienced by humans or how they interact with each other. This approach emphasizes that objects have a reality and significance beyond their interactions. The result is a de-centring of the human, which is the prerequisite for the paradigm shift in ecological thought (Willis 2).

Morton's subsequent book *Hyperobjects: Philosophy and Ecology after the End of the World* (2013), further elaborates on the concept. As objects expand and multiply, some of them become so vast as to elude human perception and cognition. They cannot even be described or understood in the ordinary spatiotemporal sense (Bricker 359). Morton defines them as "massively distributed in time and space relative to humans" (*Hyperobjects* 1). Nuclear materials, global warming, black holes, evolution, and capitalism are hyperobjects because they are so vast and pervasive that they exist on another order of magnitude. Plutonium-239, used as fuel for nuclear reactors, is an isotope, a simple object. But it is also a hyperobject, because it emits deadly radiation and has a half-life of 24,100 years, which is a time scale incomprehensible to humans. In his 2012 article "Everything We Need: Scarcity, Scale, Hyperobjects" Morton comments on the timescales of global warming:

The terrifying (500 years, 75%); the horrifying (30,000 years, 25%); the petrifying (100,000 years, 7%). In 30,000 years, let alone 100,000 years, two things will be true:

1. No one will be meaningfully related to me.
2. Everything that happens will bear the traces of my actions. ("Everything We Need," 81)

Morton uses the staggering timescales of global warming to illustrate both the enormity and the pervasiveness of hyperobjects. They are everywhere and nowhere at once: they cannot be experienced directly but they surround us and reach far into the future. Hence, hyperobjects are non-local; they cannot be directly observed, although their effects are felt by human objects. Global warming happens on a scale not directly observable to us, but the effects of global warming influence and very often threaten our daily lives. Hurricanes, heatwaves, forest fires, flash floods, rising sea levels, and melting ice are all local manifestations of "some vast entity that we are unable directly to see" (*Hyperobjects* 47). Hyperobjects also display viscosity: they stick to other entities and influence them, blurring boundaries and highlighting the entanglement of diverse elements. The feeling of heat on our skin is the effect of global warming, and this hyperobject becomes stuck to us in the form of rashes, burns, and cancer (Bricker 360). Humans cannot feel the radiation from plutonium-239, but it manifests in the form of radiation sickness and cancer. Whether we are aware of hyperobjects or not, they are there, impossible to shake off (*Hyperobjects* 35).

Hyperobjects are temporally undulated because they exist on such a massive timescale that they become impossible to hold in mind. Temporal undulation is closely related to phasing, meaning that hyperobjects "occupy a high-dimensional space that makes them impossible to see as a whole on a regular three-dimensional human scale" (Bricker 360). Morton compares this characteristic with observing an iceberg. Only the tip of an iceberg is visible above water, while the remaining 90 percent is under water. This part can be observed by moving the camera underwater, but the action then distorts the visible part of the iceberg, rendering the complete picture impossible. Similarly, we can only observe "thin slices" of hyperobjects, like hurricanes, tsunamis, or radiation sickness, but a hyperobject in its entirety evades observation.

Hyperobjects display interobjectivity, a characteristic that is closely connected to Morton's understanding of ecology and ecological thought. Morton begins *Hyperobjects* by critiquing traditional conceptions of ecology, which often depict nature as a separate entity from human society. He argues that this perspective is inadequate for addressing the complex environmental challenges of the modern world. He then introduces the concept of the "mesh," a term he uses to describe the intricate, interwoven connections between all forms of life, human and nonhuman, in a shared ecological space. The mesh underscores the inherent interconnectedness of all entities, challenging the notion of nature as an external, untouched realm. Hyperobjects constitute the mesh of cause and effect, resembling force fields: we can often detect their presence only by the footprints or cascading effects they have on other objects.

Morton's exploration of ecological thought has significant implications for human existence and sense of self. The concept of the mesh challenges the traditional idea of individuality, emphasizing the interconnectedness of all entities. Morton argues that the anthropocentric view that places humans at the centre of the universe should be abandoned. He calls for an ecological mindset that recognizes the complex, symbiotic relationships between humans and their environment, urging a shift in ethical and moral considerations. If we can recognize our interconnectedness with the environment, then our ethical considerations should extend beyond human-centric concerns to encompass the well-being of all entities within the mesh. This expanded ethical framework challenges us to consider the far-reaching consequences of our actions on a planetary scale, particularly considering pressing environmental issues such as global warming and habitat destruction.

In the second part of the book, Morton discusses the ethical, social, and political systems we need to develop to respond adequately to the time of hyperobjects. The rapid pace of environmental degradation has shown that hyperobjects create a web of interconnectedness, spanning past, present, and future, inside which all human and non-human objects float. Hyperobjects—such as oil, radiation, global warming, and Styrofoam—exist on a different spatiotemporal scale and will outlast every human being alive now by many lifetimes. Therefore, humans must adopt "an ethics of the other, an ethics based on the proximity of the stranger" (*Hyperobjects* 124) and learn to responsibly coexist with non-human objects.

The eco-concept of hyperobjects is especially relevant in the Anthropocene, an epoch characterized by human activities significantly altering the Earth's ecosystems. Global warming, often considered a quintessential hyperobject, exemplifies the ways in which human activities can have far-reaching consequences that transcend conventional spatial and temporal boundaries. Morton focuses on global warming as a man-made hyperobject, whose presence and impact are often denied, and uses it to illustrate the weaknesses of modern environmental philosophy. He argues that "sustainable capitalism" and "green policies" are not feasible, because they use the "environmentally friendly" cloak to disguise the existing economic and social systems. These systems will only continue to reproduce substances that will further expand hyperobjects, the obvious example being energy consumption and global warming. Secondly, he believes that environmentalism is still too anthropocentric and focused on preserving human objects, often at the expense of non-human objects. Finally, he claims that environmentalism is too focused on public relations, i.e., persuading people to act in environmentally friendly ways. However, Morton feels that this strategy has failed and that most people will continue to deny their role in the Anthropocene. Again, global warming is a telling example: although the effects are clear for all to see, manifested as they are by extreme weather events, melting ice, and rising sea levels, governments and nations continue to do nothing or very little to prevent or contain the activities that aggravate global warming (e.g., Morton mentions the skeptics who deny global warming based on their own local experience). Further reasoning will not help, Morton feels; "rational discourse has failed, and the affective domain must be awakened through the magic of art" (Bricker 363).

Object-oriented art and philosophy can help us transcend the limitations of language and perception by engaging with the complexities of hyperobjects. Artistic representations of hyperobjects can evoke a sense of their vastness and interconnectedness, inviting audiences to consider the implications of these entities on human existence and the environment. Object-oriented art stimulates contemplation and attunes humans to non-human hyperobjects. "For Morton, it is this dizzying contemplation that will help humans wake up to hyperobjects that have been, are already, and most definitely will continue beyond human existence" (Bricker 363). The 2023 biopic *Oppenheimer* by Christopher Nolan, for instance, presents atomic energy, nuclear materials, and radiation as hyperobjects: they are non-local, viscous, and pervasive. They shatter the world as we know it; once awakened, these forces are too vast and far-reaching for humans to comprehend or control, and yet they have massive implications on human and non-human objects. Interestingly, in his book Morton references J. Robert Oppenheimer and the Trinity test to illustrate the type of hyperobject—global warming or atomic bomb—created by humans but ultimately outside of their control. In this case, the artistic representation of the dangers of using hyperobjects as weapons may affect audiences more than any scientific or political discussion on the topic. Rather than elaborating too much on the science and ethics of harnessing atomic energy, the film uses visuals and sounds to represent its workings: fluorescent streaks of light as atoms collide and split; the blinding burst of the detonation, the eerie silence that follows, and the earth-shattering boom; the roar of the flames as cities are consumed. The film references the facts that some physicists in the 1940s hypothesized that once started, the splitting of the atoms would provoke a chain reaction that would incinerate the atmosphere; this is depicted as a powerful visual of the planet engulfed by a tsunami of fire.

The end of the world will not be a fiery future cataclysm though; as Morton argues, it has already occurred for all who understand that, in the age of hyperobjects, there can be no clear-cut distinctions between "human" and "Nature." We are all entangled in the mesh: human, non-human, space, and time. *Hyperobjects* argue in favour of ecocentrism, an eco-concept that became prevalent in environmental ethics in the past 50 years. Like object-oriented ontology and speculative realism, ecocentrism rejects anthropocentrism and promotes the equality of human and non-human entities. Moreover, ecocentrism sees the roots of the environmental crisis in anthropocentrism and in people-centred policies and practices (Angus, "Anthropocentrism vs. Ecocentrism"). The book's subtitle *Philosophy and Ecology After the End of the World* is meant to wake policymakers to the reality of hyperobjects. The usual reaction to the vastness and incomprehensibility of hyperobjects is to withdraw from them or displace them. However, nuclear waste does not stop emitting deadly radiation just because it is buried inside a mountain. In this way, the environmental costs are only shifted onto future generations.

The concept of hyperobjects has generated both support and criticism within academia. Morton's ideas have been lauded for their capacity to reshape the anthropocentric perception of reality and to emphasize the urgency of addressing global ecological challenges. Many critics claim that Morton's hyperobjects help us understand the interconnectedness of the human and the non-human and emphasize the importance of the concept for modern ecological thought (cf. Bricker 364; Cannella 10–12; Welling 3–4). Hyperobjects provide a framework for understanding phenomena like climate change and global warming, shedding light on the complex, interconnected systems at play. Researchers from a variety of disciplines, such as environmental philosophy, climatology, architecture, and archaeology, have proposed a number of entities that could be qualified as hyperobjects that possess agency. For instance, archaeology scholar Peter B. Campbell argues that the sea is not a landscape but a vast spatial and temporal entity with an agency of its own ("The Sea as a Hyperobject"); in "The Anthropocene, hyperobjects and the archaeology of the future past" he claims that the entire archaeological record is a hyperobject, "human residue" consisting of not only artifacts and sites but also of eco-facts (e.g. greenhouse gasses), hyperfacts (water, radioactivity) and digital facts (data generated by the Internet of Things), all of which will impact the future and become archaeological data (1326).

Hyperobjects have also influenced artists and writers concerned with environmental issues and ecological thought. It has been suggested that this eco-concept be used to help environmental educators raise ecological awareness in a time of crisis (Bazzul 213; Bradley 163). The 2015 book *Art in the Anthropocene: Encounters Among Aesthetics, Politics, Environments and Epistemologies* edited by Heather Davis and Etienne Turpin explores how art can address hyperobjects and the environmental challenges they represent. The artistic representations of hyperobjects help bridge the gap between abstract philosophical concepts and lived human experiences, fostering a deeper connection to ecological issues. In literature, the eco-concept has been applied to new critical readings focused on decentring human subjectivity (Kortekallio 59–60) and to formulating new "eco-poetics that encourage readers to think in terms of inescapable ecological interrelatedness—what Morton terms the mesh" (Keller 2–3).

The eco-concept of hyperobjects has also provoked criticism. Given the non-local and temporally undulating nature of hyperobjects, some argue that hyperobjects risk becoming abstract and detached from concrete experiences, potentially limiting their practical applicability (Bricker 365; Welling 3–4). Critics have also noted how hyperobjects and object-oriented ontology profoundly reshape the human-centred perception of reality: the result is the de-centring of the human which poses problems for many disciplines that focus on humans and human activity (Willis 2). Furthermore, the inherent complexity of hyperobjects may render them inaccessible to individuals who lack a philosophical background. Finally, some scholars note that Morton does not provide sufficient evidence that the concept of hyperobjects and the awareness of their existence will change environmental attitudes (Bricker 366, Welling 4). Hyperobjects have shattered the human-centred ontology, but it is not clear how we should proceed: there is no advice on how to stop global warming. Morton himself seems to believe that reshaping our human-centred perception and recognizing the reality of hyperobjects is an important step forward.

An interesting take on hyperobjects comes from Frantzen and Bjering (2) who develop a new theoretical concept of the "hyperabject," combining Morton's concept of hyperobjects and Julia Kristeva's theory of abjection. This concept includes both a critique of and correction to Morton's ecological thought, as well as speculative realism and object-oriented ontology. The authors agree with Morton about the reality of hyperobjects but argue against the notion that all of them have agency. They warn against too-flat ontologies like object-oriented ontology that claims that everything exists equally, i.e., nothing has a special status (3). Treating global pollution and waste as hyperobjects obfuscates the urgency of this problem and the issue of responsibility and agency. Environmental pollution is the "hyperabject"—waste produced by various human activities that cannot go away; it cycles endlessly through the biosphere and the human body (microplastic is a good example). It is characterized by inertness and lack of agency. Frantzen and Bjering criticize the lack of a political dimension in Morton's ecological thought: by emphasizing the interconnectedness and the mesh, it fails to recognize the true cause of the climate crises, which is capitalism with its "machine of extraction, exhaustion and extinction" (15). The hyperobject, they argue, does not allow us to pose the proper questions to capitalism and the ecological destruction wrought by the ongoing expansion and exhaustion of all possible resources, nor any real answers to it (4).

The eco-concept of hyperobjects presents intriguing possibilities for reshaping our worldview and addressing ecological crises. By recognizing the pervasive influence of hyperobjects, policymakers, scientists, and activists might adopt more comprehensive strategies for addressing ecological concerns. Scholars from various disciplines, such as environmental philosophy, climate activism, education, architecture, and archaeology, have addressed the topics of hyperobjects in their research. Moreover, the concept's influence on cultural and artistic expression is likely to grow. Artists and writers will continue to employ innovative techniques to convey the complexity and interconnectedness of hyperobjects. This artistic engagement could raise public awareness of ecological issues. However, it is crucial to find ways to make the concept more accessible and applicable across diverse disciplines and audiences. Additionally, continued dialogue and collaboration between philosophers, scientists, artists, and policymakers will be crucial to refine and expand the concept's utility in addressing the ever-evolving challenges of our interconnected world.

This eco-concept has been useful in raising awareness about the far-reaching and often devastating impact of phenomena such as global warming, radiation, nuclear materials, and capitalism. As a result, policymakers, scientists, and activists might adopt more comprehensive strategies for addressing ecological concerns. Scholars from various disciplines, such as environmental philosophy, climatology, education, architecture, and archaeology, have addressed the topics of hyperobjects in their research. Moreover, the concept's influence on cultural and artistic expression is likely to grow. Artists and writers will continue to employ innovative techniques to convey the complexity and interconnectedness of hyperobjects. This artistic engagement could raise further public awareness of ecological issues.

Hyperobjects and object-oriented ontology have played an important role in environmental ethics by decentring the human, emphasizing the interconnectedness of ecological systems, and acknowledging the limitations of human perception and representation. However, there is an inherent risk in de-centring human agency, and that is the question of responsibility. If every object is equal and has an agency, who is responsible for pollution, environmental degradation, and oil spills? Hyperobjects must include a political dimension and accountability in order to expand the utility of this eco-concept in confronting environmental challenges.

REFERENCES

Angus, Ian. "Anthropocentrism Versus Ecocentrism: Notes on a False Dichotomy." *International Socialism,* Issue 171, July 23, 2021, www.isj.org.uk/anthropocentrism-versus-ecocentrism/. Accessed September 14, 2023.

Bazzul, Jesse. "Hyperobjects, Media, and Assemblages of Collective Living: Playing with Ontology as Environmental Education." *Australian Journal of Environmental*

Education, Volume 35, Issue 3, November 2019, pp. 213–21 https://doi.org/10.1017/aee.2019.20. Accessed September 14, 2023.

Bradley, Joff P. N. "Transcendental Monsters, Animism and the Critique of Hyperobjects." *Australian Journal of Environmental Education,* Volume 35, Issue 3, November 2019, pp. 163–172, https://doi.org/10.1017/aee.2019.15. Accessed August 24, 2023.

Bricker, Brett. "Hyperobjects: Philosophy and Ecology After the End of the World by Timothy Morton." *Philosophy & Rhetoric*, August 31, 2015, Volume 48, Issue 3, pp. 359–65, https://doi.org/10.5325/philrhet.48.3.0359

Campbell, Peter B. "The Anthropocene, Hyperobjects and the Archaeology of the Future Past." *Antiquity,* Volume 95, Issue 383, 2021, pp. 1315–30, doi: 10.15184/aqy.2021.116.

———. "The Sea as a Hyperobject: Moving Beyond Maritime Cultural Landscapes." *Journal of Eastern Mediterranean Archaeology & Heritage Studies*, Volume 8, Issue 3–4, 2020, pp. 207–25. https://doi.org/10.5325/jeasmedarcherstu.8.3–4.0207. Accessed September 14, 2023

Cannella, Megan E. "Review of Hyperobjects: Philosophy and Ecology after the End of the World." *Journal of Ecocriticism,* Volume 6, Issue 2, July 2014, pp. 10–13, https://ojs.unbc.ca/index.php/joe/article/viewFile/620/506. Accessed August 24, 2023.

Frantzen, Mikkel Krause and Jens Bjering. "Ecology, Capitalism and Waste: From Hyperobject to Hyperabject." *Theory, Culture & Society,* Volume 37, Issue 6, pp. 87–109, https://doi.org/10.1177/0263276420925541

Kortekallio, Kaisa. "Becoming-Instrument: Thinking with Jeff VanderMeer's *Annihilation* and Timothy Morton's *Hyperobjects*." In *Reconfiguring Human, Nonhuman and Posthuman in Literature and Culture*. Edited by Sanna Karkulehto, Aino-Kaisa Koistinen & Essi Varis. Routledge, 2019, pp. 57–75, https://helda.helsinki.fi/server/api/core/bitstreams/65e7b0cd-7e6d-49f4-a53c-44cfd95b1323/content. Accessed September 15, 2023.

Keller, Lynn. "The Ecopoetics of Hyperobjects: Evelyn Reilly's *Styrofoam*." *Interdisciplinary Studies in Literature and Environment,* Volume 22, Issue 4, 2015, pp. 846–71, https://doi.org/10.1093/isle/isv072. Accessed September 14, 2023.

Morton, Timothy. "Everything We Need: Scarcity, Scale, Hyperobjects." *Architectural Design* Volume 82, Issue 4, 2012, pp. 78–81, https://doi.org/10.1002/ad.1433. Accessed August 24, 2023.

———. *Hyperobjects: Philosophy and Ecology After the End of the World.* University of Minnesota Press, 2013.

Welling, Bart H. "A New Era of Scholarship"? *Hyperobjects* and Hyperobjects. Review of *Hyperobjects: Philosophy and Ecology After the End of the World,* by Timothy Morton." *The Goose*, Volume 13, Issue 2, Art. 27, 2015, https://scholars.wlu.ca/thegoose/vol13/iss2/27. Accessed August 25, 2023.

Willis, Kimberley. "JCA Book Reviews: *Hyperobjects: Philosophy and Ecology after the End of the World,* by Timothy Morton." Equinox Publishing, February 23, 2015, http://www.equinoxpub.com/home/jca-book-reviews-hyperobjects-philosophy-ecology-end-world-timothy-morton/. Accessed August 25, 2023.

PART II

Exploring Ecocriticisms

Chapter 12

Restoration Ecocriticism

From Habitat Destruction to Hands-on Action

Ufuk Özdağ

Restoration ecocriticism that I set forth more than a decade ago is now timely as 2021–2030 is announced as the UN Decade on Ecosystem Restoration. "There has never been a more urgent need to revive damaged ecosystems than now," reads a headline in their website.[1] Edward O. Wilson's Half-Earth Project, devoting half the surface of the earth to nature, provides a useful road map with which an ecologically sustainable relationship with the ecosystems can be implemented.[2] I first mentioned the need for a "restoration ecocriticism" in my "An Essay on Ecocriticism in the 'Century of Restoring the Earth'" (2009), voicing concern about a lack of a restoration component within ecocriticism, the field of study that mobilized the literary scholar for the last three decades to heal the earth.[3] During this time span, ecocritics responded to a range of environmental challenges, but did not pay attention enough to the call of the restoration movement that conservationists have been engaging for many years.[4] As a response to increased urbanization and industrialization that have severe consequences for how lands are used, Aldo Leopold's biographer and conservation biologist Curt Meine once stated,

> If we equate conservation with a simple and static notion of preservation, and regard restoration as a separate undertaking that seeks to reestablish lost or degraded ecological qualities, then the idea of "novel ecosystems" obviously presents fundamental challenges to the very notion of restoration. If, however, we regard conservation as encompassing varied and dynamic relationships between humans and nature, and ecological restoration as one expression of those changing relationships, then restoration remains vital and relevant. Thus,

> the idea of novelty tests the viability of ecological restoration, and restoration conversely tests the practical relevance of novelty to conservation. (Meine, "Restoration and Novel Ecosystems" 218)

How can this potential ecocritical school, restoration ecocriticism, help communities around the world to develop a sense of stewardship and responsibility for land and its nonhuman inhabitants with engagements in hands-on restoration?[5] I've been encouraging the growth of this new approach so that implementing collaborative and dynamic projects that build awareness, appreciation, and responsibility for conserving and creating rich ecological landscapes around the world becomes habitual.[6]

Recently, in "Feeling Like the Colorado River: Laying the Groundwork of Restoration Ecocriticism" (2022), I provided a definition for restoration ecocriticism, and outlined the task of the ecocritic:

> Restoration ecocriticism is the ecocritical study of literary and cultural texts that explore or inspire individual or collaborative community restoration efforts in the degraded lands/waters/marine environments, most often caused by anthropogenic activities.[7]

I also listed the three distinctive features of this new ecocritical school, saying that hope will replace despair in ecocritical scholarship in the rush to reverse the fate of degraded local lands, that natural history will become an integral part of ecocritical studies with emphasis on lands before and after anthropogenic damage, and that the ecocritic will connect individual or collaborative community restoration projects to sustainability in the local areas for community well-being (11). The goal of restoration ecocriticism, then, is to highlight literary texts that will support an emerging ecological citizenship dedicated to restoring degraded lands, as well as to bring awareness about the actual land healing efforts in the most hands-on way. Aldo Leopold's life and legacy is important in this context. The father of ecological restoration, Leopold is widely acknowledged for the restoration of his shack property in Baraboo, Wisconsin, in the 1930s, but his grand legacy is that of helping restore the Coon Valley, the restoration of a wide expanse of land in Wisconsin, a first in world environmental history that proved that the health of an ecosystem can be reintroduced through collective collaborative action (Meine).[8] In this region, land was devastated to such an extent that gullies were formed deep and wide, salmon was depleted in the Coon Creek, and sustainability was totally gone. It was absolutely necessary to lead lives in accordance with the ecosystem laws that had shaped the lands across millennia. As a matter of fact, in "Marshland Elegy" in *A Sand County Almanac*, the essay that led to the restoration of many wetlands in Wisconsin,[9] Leopold pointed at the geological processes, how the "crane marsh" was formed over millennia in the Baraboo landscapes,[10] and that altering the natural processes ultimately backfires "in the march of aeons." Leopold stated, "[A] crane marsh holds a paleontological patent of nobility, won in the march of aeons," and he urged its restoration saying, "the sadness discernible in some marshes arises, perhaps, from their once having harbored cranes. Now they stand humbled, adrift in history" (97). Leopold's land restoration legacy, then, is grounded in the knowledge of the land shaped by geological forces over millennia. In line with these ideas, restoration ecocriticism is the literary reflection of restoring lands to their original state, of ecological necessities, of reverence to biological life processes, of respect to geological forces. Continuing the Leopold family legacy, Scott Freeman, in his nature writing book *Saving Tarboo Creek* (2018) significantly stated, "In restoration is the preservation of the world" (97).[11]

Environmental narratives that trigger landscape-scale restoration abound from around the world. Turkey's legendary environmental writer, Yaşar Kemal's novels in his *Akçasazın Ağaları* series [*The Lords of Akchasaz*], elegies for the once rich lands, facilitate such interest. I pick this series as there is urgency to restore wide expanses of lands that had seen massive degradation by the tragic drainage of the watery scapes more than seventy years ago that Yaşar Kemal chronicled in the 1970s, the very lands that were shaken by the recent earthquakes in Turkey.

On February 6, 2023, the world became silent, listening to devastating news from Turkey, one of the world's most active earthquake zones. In the 7.8 and 7.6 magnitude earthquakes that struck Kahramanmaraş, eleven cities spreading out to 160 km were affected, and tens of thousands of buildings collapsed.[12] In the "calamity of the century," the city of Antakya in the Hatay region, for instance, was totally wiped out. The human losses are indescribable; the cultural losses will be much debated. Earthquakes are natural phenomena, and yet, a destruction of this scale is not natural as in the earthquake zone, ecosystem laws and geological forces were disregarded for more than a century. In the Hatay region and the environs, for instance, there was once a mania for draining the lakes, ponds, and wetlands for decades. The new settlements spread out to plains, with reckless construction of apartment buildings on top of alluvial soils. The earthquake brought misery of an unprecedented magnitude. Can the literature profession alleviate such destruction? Many years ago, when I visited the Hatay area for two summers to take part in outdoor nature education, I was stunned to learn about the relentless drainage of the Amik Lake (Lake Antioch) in this area, once located north-east of the ancient city of Antioch (Antakya), once the breeding ground of hundreds of thousands of waterbirds on the African Eurasian flyway, a habitat shaped by the living Earth over millions of years. In an essay written in dismay, I stated that restoration of this area is the most urgent environmental need of the country ("Keeping Alive the Memory of the Amik" 2011).[13] Based on reportages with the local peoples, I stated that the literature scholar should draw attention to human wrongs, to create awareness, to restore lands and freshwater systems. With grief for the lost paradise of Amik, I had set an Amik elegy, inspired by Leopold's "Marshland Elegy," and voiced my urgency:

> Embracing what I would like to call a restoration ecocriticism, i.e., land restoration as a topic within ecocritical studies, then, in the face of mounting environmental problems, continuing degradation, and loss of sustainability, may contribute to the pioneering work done in many local bioregions. ("Keeping Alive the Memory of the Amik" 121)

Now, at the earthquake ruined Amik area as well as the entire earthquake zone, there is a need for "thinking communities" to restore the lakes and wetlands that had seen massive degradation over the decades, for a sustainable future. Lands are crying out for their watery scapes prior to 1950s, once biodiversity hotspots in the immensity of deep time.

A CASE STUDY FOR RESTORATION ECOCRITICISM

What does restoration ecocriticism mean? I propose that the ecocritic proceeds from an actual case of profound anthropogenic change in the land, then urges restoration work by highlighting a literary/cultural text to draw attention to the case. In other words, the ecocritic uses narratives of human impacted devastation to repair and bring back the land's vital processes. The ways in which the narratives keep the memory of diminished landscapes, waterscapes alive, and draw attention to before/after states of the degraded ecosystems will be crucial in the analysis.[14]

Reading the two-volume novels in *The Lords of Akchasaz*, the stories of human impacted land devastations in the Çukurova region, narrated by Yaşar Kemal in the 1970s, tells us why the recent earthquakes brought so much misery. The setting in *The Lords of Akchasaz* series is the vast area hit by the earthquakes, where wetlands were once drained to open up new croplands, much like the drainage of the Amik Lake and surrounding wetlands in Hatay, and became one of the centers of unnatural destruction.[15] The novels in the series are masterpieces steeped in history, that will play a role in the restoration of the devastated lands, wetlands, and watersheds of the region.[16]

The time span in *The Lords of Akchasaz* series is the advent of mechanization in agriculture, and the arrival of thousands of trucks to the South for intensive agriculture in the 1950s. The setting is the wetlands, to the south of Anavarza Antique city, which Kemal points out in *Yusufçuk Yusuf* as 30.000 acres (56). The storyline is about the fierce quarrels of new generation farmers in the Savrun Stream region to seize land from the Akchasaz marshland. Once the habitat of numerous birds and wildlife, the area saw massive degradation in the 1950s.[17]

Environmental History of Akchasaz and its Devastation

The environmental history of Akchasaz marshes reveals that once the area to the south of the Anavarza Antique city was used for intense rice cultivation. The first novel in the series *Demirciler Çarşısı Cinayeti* [*Murder in the Ironsmiths Market*] reveals the disputes among the landlords.[18] A "sharp-witted" landlord, Amber Ağa, opens up a rice arch in 1874, and makes a fortune out of rice cultivation. In a particularly wet season, the Savrun Stream changes its course, flooding thousands of acres of land. In time, a vast expanse of marshland, Akchasaz, fed by the Savrun Stream comes into being which falls into the hands of rice farmers. In *Murder in the Ironsmiths Market,* the narrator reveals that, over the years Akchasaz got flooded and became even more expansive, reaching Amberinarkı in the North, Kesikkeli, Endele, and Ceyhan in the South (137) (Figure 12.1). As the narrator reveals "a dizzying appetite for land" comes into being (168). In the later years, as narrated in *Murder in the Ironsmiths Market,* Savrun Stream starts drying up, Akchasaz becomes devoid of its main water source, the Savrun. Because of excessive water use by rice farmers, Akchasaz wetlands start drying up. The muddy lands of Akchasaz fall into the hands of new generation farmers. Each year it dries up some more, and each year villagers, landlords crowd onto the drying Akchasaz, they plunder the wetland, opening up ditches and drainage canals (137). The narrator in *Murder in the Ironsmiths Market* reveals that poor people take out "thousands of acres" from Akchasaz, and build large farms (162), depleting the biodiversity and wildlife. Borrowing words from Leopold's "Marshland Elegy," they "envisaged farms not only around, but in the marsh. An epidemic of ditch-digging and land booming set in" (Leopold 100). Thus, the pillage of Savrun waters, followed by the draining of the wetlands by the Kadirli landlords characterizes the environmental history of the region.[19] In time, Kadirli landlords resume rice cultivation, dug up additional arches from the Savrun Stream, and they build even vaster marshlands next to Akchasaz. It is such massive rice cultivation in the plains that water wars come into being among the rice farmers (*Murder in the Ironsmiths Market,* 137). This pillage of the Savrun Stream leads to the devastation of the region's water regime in the 1950s to which Yaşar Kemal responds with

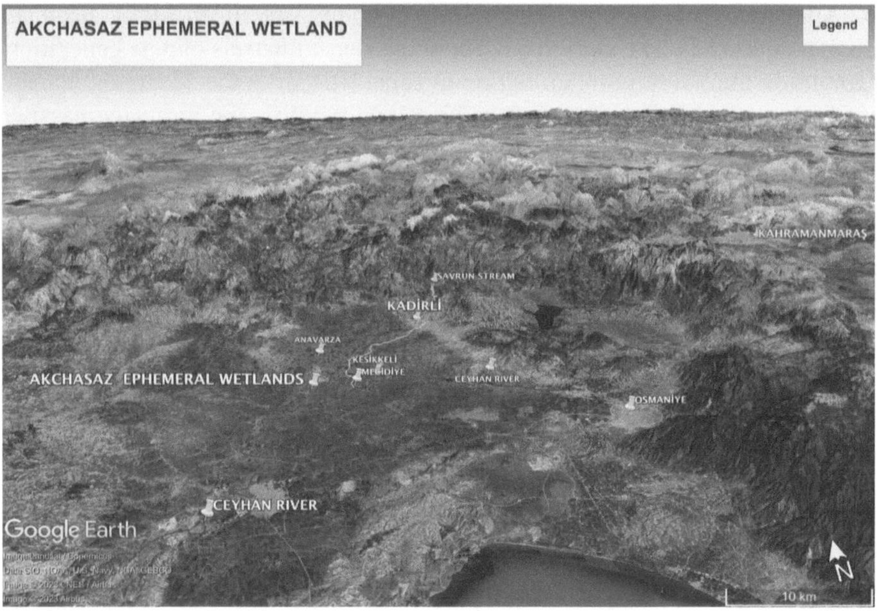

Figure 12.1. The location of Akchasaz ephemeral wetlands. *Source*: Created by Dr. Cemal Saydam

his monumental *The Lords of Akchasaz,* the two-volume series published in 1974 and 1975.[20]

The epic novels of *The Lords of Akchasaz* series, *Murder in the Ironsmiths Market* and *Yusufçuk Yusuf* [Turtledove Yusuf] each over 600 pages, reveals Yaşar Kemal's concern and grief for an ecosystem that was unique in the world. The biodiversity that Kemal narrates with a botanist's eyes, the watershed filled with eagles, pink pelicans, geese, ducks, and numerous other migratory birds shows Yaşar Kemal's belief in the wetlands' right to live.[21] Çukurova was once a paradise, with numerous waterbird species and reed beds, until a time came when trenches and canals left not "a single drop of water." Yaşar Kemal narrates the tragic events as:

> The Akchasaz marshland was located on the Anavarza Plain. There were 17–18 marshes in Çukurova. Forest, reed, swamp. . . . This lasted until 1950. It was a slow process. All the marshlands were dried up. There used to be oaks. . . . Now there is nothing. Not a single tree remained along the Mediterranean. Historians say that even in the nineteenth century there used to be a large forest . . . this was how nature was seen way before the transition into capitalism as part of this feudal system: swamps, gorse, woodlands, reeds . . . there is a terrible lack of natural vegetation and I have caught up with this natural cover. Hundreds, thousands of tractors entered Çukurova, it happened so quickly . . . an

incredible destruction of nature. Çukurova was transformed into an agricultural desert. . . . Do you ever see an antelope in Çukurova? Thirty thousand antelopes used to roam around Çukurova in February. (Andaç, *Söz Uçar Yazı Kalır,* 253)

Yaşar Kemal narrated the collapse of Akchasaz ecosystem in a lengthy reportage in 2003, as well. Saddened he stated, "We have wasted that legacy . . . despite knowing it was our life blood, we denied it."[22]

The Lords of Akchasaz is an Elegy to Wetlands

The water disputes for the Savrun Stream that fed the Akchasaz, the advent of agricultural machinery and the arrival of thousands of trucks to the region in the 1950s, the new generation farmers seizing the wetlands, and ensuing drainage works are in the center of *The Lords of Akchasaz*, the Turkish marshland elegy.[23] Those who remember the olden times of the region are grief-stricken narrating its once celebrated biodiversity. They narrate that the purest waters of the Savrun and extensive marshlands were refuge to numerous waterbird species, that healthy strands of aquatic plants played a vital role in the ecology of the expansive wetlands, that all life was harmonious and healthy. The wonderlands of biodiversity in which Yaşar Kemal grew up and the pillage of the Savrun Stream region bare similarities to the pillage of the Amik Lake and its surrounding wetlands, the healthy landscapes formed by tectonic forces over millions of years.[24] This freshwater lake on the flyway used by hundreds of thousands of migratory birds annually was drained in the 1950s, 1960s, and 1970s, and is now re-emerging following the two major earthquakes that cracked open the earth's surface, on February 6, 2023. The publication of Yaşar Kemal's *Murder in the Ironsmiths Market* and *Yusufçuk Yusuf* coincides with the times when the Amik Lake was totally gone. Yaşar Kemal narrates not the Amik tragedy, but the nearby Akchasaz, and its ecosystem collapse, the lands where he was born and raised. For Kemal, the Akchasaz marshland was intrinsically valuable. The new generation farmers who lacked ecological conscience used every means to take out land from the Akchasaz with newly owned giant machinery:

The villagers were attacking from all directions as everyone was in a battle to occupy land by digging small tunnels from the swamp next to their field, filling the shallows with earth. The greed to occupy the land was a feverish disease. Everyone was attacking the swamp from one side, hundreds of small canals, ditches, arches descended from the reeds to the dried bed of the swamp, the water of the swamp was constantly draining, leaving national lands on the shores. The lands of the peasants who had fields along the coastal region were expanding day by day. Some of the villagers had already owned large ranches. (*Yusufçuk Yusuf* 38)

As narrated in *The Lords of Akchasaz*, the wetlands that once harbored numerous waterbird species, were drained and depleted in a short time span, wiping Akchasaz from history. It is important to note that the drainage completely destroyed local biodiversity and devastated local economies. Migratory birds are now seldom seen in this area. Many freshwater plant species have become extinct, and fishing has come to a tragic end. To remember the words of Ozaner, "[a]n area determined as a wetland by techtonics cannot be used as a cropland, for the area will always cry out for its water."[25]

RESTORATION ECOCRITICISM IN THE UN DECADE ON ECOSYSTEM RESTORATION

Now the crucial question in the UN Decade on Ecosystem Restoration is how to motivate restoration work at the landscape-scale for a more sustainable future. Such efforts will affect large areas of the earth and change the lives of millions of people. Human impacted land degradation from the recent earthquake zones as reflected in *The Lords of Akchasaz* series reveals that there is need for successful hands-on restoration stories from around the world. This will bring hope. A recent report by the United Nations Environment Programme and the IUCN reveals that "202,467 protected areas currently exist, accounting for 14.7% of Earth's land." Given the scale of degraded lands, this is only halfway, and requires laborious work to reach Edward O. Wilson's Half-Earth Project (Krupnick and Knowlton 336). Restoration ecocriticism will motivate communities to take an active role, to underline the need for restoring degraded lands. As Krupnick and Knowlton state, we need to bring to the forefront the success stories of restoration to motivate communities and have them join collective efforts (337). Restoration ecocriticism promotes collective collaborative action, but more importantly, it highlights success stories of land and species restoration to initiate new work in the local lands. I remember my excitement reading Freeman House's *Totem Salmon* and Scott Freeman's *Saving Tarboo Creek*, two of these treasures of collective collaborative action, guiding us for our future restoration work. Influential literary texts such as *The Lords of Akchasaz,* crucial for bringing back land health to once earthly harmonies, can trigger flagship initiatives such as the projects on the website of the UN Decade on Ecosystem Restoration.[26] Land restoration is no longer a dream around the world.[27] In the second half of the twentieth century in Turkey, numerous wetlands were drained, a significant portion is on the recent earthquake zone, the Akchasaz area included. With impressive environmental history in its background, *The Lords of Akchasaz* series will inspire action and make meaningful change to bring land health to the region.

The aftershocks have not yet ceased in the world's most active earthquake zone, reminding us to hear the call of wide expanses of degraded lands to revive the lost ecosystem harmonies. There is an urgency to convince decision-makers, and to initiate collaboration across the public, private and governmental sectors to transform the watersheds. Already local peoples in this area are negotiating ways to bring back their lakes and wetlands, and ready for hands-on restoration, knowing that this will restore human communities, as well. The restoration of the Amik watershed is widely discussed as the underground water levels have risen following the earthquake, tectonics cooperating for land healing efforts. Literature scholars are needed more than ever for communicating skills and knowledge in healing the lands, for disseminating restoration research and practices from around the world, for triggering grassroots restoration activities.

From its inception, ecocritical theory recited human bodies and the earth body as one integral whole. Cheryll Glotfelty, in her seminal essay, "Literary Studies in an Age of Environmental Crisis," proclaimed, "all ecological criticism shares the fundamental premise that human culture is connected to the physical world, affecting it and affected by it" (xix). Now, decades later, ecocriticism needs to take up hands-on action, bringing reciprocity to the healings, as this could also address ecocriticism's methodological questions. Restoration ecocriticism, as a vital ecotheory to revisit literary and cultural texts, may become one of the means to raise awareness, to bind up wounds, and to mobilize community involvement for landscape-scale Earth repair.

NOTES

1. See UN Decade on Ecosystem Restoration, https://www.decadeonrestoration.org/.

2. See https://eowilsonfoundation.org/what-is-the-half-earth-project/.

3. See Özdağ, "An Essay on Ecocriticism in the Century of Restoring the Earth," 140.

4. See Clewell and Aronson, "Motivations for the Restoration of Ecosystems," 421.

5. See my words, "In an article I published in 2009, I coined the term 'restoration ecocriticism.' My decision was prompted by a resolution from the delegates of the 6th World Wilderness Congress (Bangalore, India, 1998) who called on the United Nations to declare the twenty-first century as 'The Century of Restoring the Earth.' Now I wish to introduce how we might utilize texts on restoration, or texts that inspire restoration, both fiction and nonfiction, to protect and to restore our wounded lands and waters, to help initiate ecological restoration in some degraded landscapes around the globe. This may help to turn the tide and reverse the Anthropocene. In view of

the magnitude of current environmental devastations, embracing a restoration ecocriticism has become more than an urgency for the environmental humanities scholar. Heartfelt narratives of restoration, in fiction and nonfiction, may help spread the significance of the ecological restoration movement to diverse communities" (Özdağ, "Feeling Like the Colorado River" 3–4).

6. See also Özdağ, "'Evrim Orkestrasının Trompeti' Turnalar: Bir Restorasyon Çevreci Eleştiri Uygulaması" ["'Cranes Are the Trumpets in the Orchestra of Evolution': A Restoration Ecocriticism Approach"], 2019.

7. See Özdağ, "Feeling Like the Colorado River," 11. In "Feeling Like the Colorado River: Laying the Groundwork of Restoration Ecocriticism" (2022), highlighting the geologic significance of Glen Canyon, I stated that Edward Abbey made a plea for its restoration in his *Desert Solitaire.*

8. See Meine, "Conservation and Continuity." *Minding Nature*, vol. 3, no. 2 (2010), 28–34.

9. For an extensive study of "Marshland Elegy," see Meine, "Giving Voice to Concern" in *Correction Lines,* 132–147. For Leopold's restoration legacy, see Zedler, "The Continuing Challenge of Restoration," 116–126.

10. Leopold explains the geologic and glacial processes of the formation of the lake saying, "the shorelines of this old lake are still visible; its bottom is the bottom of the great marsh" (98).

11. The illustrations in *Saving Tarboo Creek* belong to Freeman's wife, Susan Leopold Freeman, granddaughter of Aldo Leopold.

12. In this deadly earthquake, there was widespread damage in an area of about 350,000 square kilometers.

13. See "Keeping Alive the Memory of the Amik." Following two visits in 2008 and 2009 to Hatay, Antakya, to take part in TÜBİTAK coordinated nature education projects, I had chronicled the draining of the Amik Lake and surrounding wetlands in the 1950s, 1960s, 1970s. Also see the video on "Voices of Green Fire: Fellow Voyagers," in which I state the urgency for the restoration of the Amik, https://www.youtube.com/watch?v=VUeipmrpeXc.

14. In "Feeling Like the Colorado River: Laying the Groundwork of Restoration Ecocriticism," I used the controversial Glen Canyon Dam in the upper Colorado region, and I listed the ecocritical questions to be directed at the related texts (12).

15. The reference is to deep alluvial soils that affected the impact of the earthquakes in the area.

16. The two novels in *The Lords of Akchasaz* series are *Demirciler Çarşısı Cinayeti* and *Yusufçuk Yusuf.* Yaşar Kemal passed away in 2015, unable to write the third novel, *Anavarza.*

17. The Akchasaz marshland was fed by the Savrun Stream, whose banks Yaşar Kemal roamed during his childhood years. Kemal recalls the Savrun Stream banks as an ocean of purple marjoram flowers that he enjoyed all day long (*Yaşar Kemal on His Life and Art,* 25–26).

18. See chapter 11 of *Murder in the Ironsmiths Market*. All translations into English from *Demirciler Çarşısı Cinayeti* and *Yusufçuk Yusuf* are mine.

19. As Yaşar Kemal points out, Kadirli landlords once dug up numerous irrigation ditches and canals from the Savrun Stream.

20. Kemal stated, "Rice farmers have eyes bigger than their stomack. They never calculate how much of the Savrun waters would irrigate a rice field. The waters do not suffice the sown field. Once that happens, quarrels follow . . . stealing water, gunfight, and murders for waters follow" (*Bir Bulut Kaynıyor*, 2003).

21. Yaşar Kemal's various conversations on Akchasaz wetlands have become more important now as there is awareness of the value of wetlands.

22. See Yaşar Kemal, "Yaşar Kemal Anadolu Doğasını Yeşil Atlas'a Anlatmıştı" ["Yaşar Kemal Narrated Anatolian Lands to Yeşil Atlas"].

23. The allusion is to Aldo Leopold's "Marshland Elegy" in which Leopold narrates the draining of a precious marshland by the "new overlords" (99), and its eventual restoration.

24. For my publications on the drainage of the Amik Lake and surrounding wetlands, see "Keeping Alive the Memory of the Amik" 2011, and *Çevreci Eleştiriye Giriş: Doğa, Kültür, Edebiyat.* Ürün Yayınları, 2014.

25. See Ozaner, "Interview with Sancar Ozaner by Ismail Zubari," October 17, 2006.

26. For impressive restoration projects, see https://www.decadeonrestoration.org/.

27. See Reid and Aronson, "Ecological Restoration in a Changing Biosphere" 185–187. An impressive example is the restoration of Everglades in the United States. See Comprehensive Everglades Restoration Plan (CERP) https://www.nps.gov/ever/learn/nature/cerp.htm

REFERENCES

Andaç, Feridun. *Söz Uçar Yazı Kalır: Yüzyılın Son Tanıkları 1.* Can Yayınları, 2001.

Clewell, Andre. F. and James Aronson. "Motivations for the Restoration of Ecosystems." *Conservation Biology*, vol. 20, no. 2, 2006, pp. 420–428.

Freeman, Scott. *Saving Tarboo Creek: One Family's Quest to Heal the Land.* Timber Press, 2018.

Glotfelty, Cheryll. "Literary Studies in an Age of Environmental Crisis." *The Ecocriticism Reader: Landmarks in Literary Ecology.* The University of Gorgia Press, 1996.

Krupnick, Gary and Nancy Knowlton. "Earth Optimism: Success Stories in Plant Conservation." *Annals of the Missouri Botanical Garden,* vol. 102, no. 2, 2017, pp. 331–340.

Leopold, Aldo. *A Sand County Almanac.* 1949. Oxford University Press, 1968.

Meine, Curt. "Conservation and Continuity." *Minding Nature,* vol. 3, no. 2, 2010, pp. 28–34.

———. "Giving Voice to Concern" in *Correction Lines: Essays on Land, Leopold, and Conservation.* Island Press, 2004.

———. "Restoration and Novel Ecosystems: Priority or Paradox?" *Annals of the Missouri Botanical Garden*, vol. 102, no. 2, 2017, pp. 217–226.

Özdağ, Ufuk. "An Essay on Ecocriticism in the 'Century of Restoring the Earth.'" *Journal of American Studies of Turkey*, no. 30, 2009, pp. 125–142.

———. "'Evrim Orkestrasının Trompeti' Turnalar: Bir Restorasyon Çevreci Eleştiri Uygulaması" ["'Cranes Are the Trumpets in the Orchestra of Evolution': A Restoration Ecocriticism Approach"]. *Anadolu Turnaları: Biyoloji, Kültür, Koruma,* edited by. Ufuk Özdağ and Gonca Gökalp Alpaslan. Ürün Yayınları, 2019, pp. 137–149.

———. "Keeping Alive the Memory of the Amik: Environmental Aesthetics and Land Restoration." *The Future of Ecocriticism: New Horizons,* edited by S. Opperman, U. Özdağ, N. Özkan, and S. Slovic. Cambridge Scholars Publishing, 2011, pp. 118–135.

Ozaner, Sancar. "Interview with Sancar Ozaner by Ismail Zubari," October 17, 2006.

Reid, J. Leighton and James Aronson. "Ecological Restoration in a Changing Biosphere." *Annals of the Missouri Botanical Garden,* vol. 102, no. 2, 2017, pp. 185–187.

Yaşar Kemal. *Akçasazın Ağaları 1, Demirciler Çarşısı Cinayeti* [*Murder in the Ironsmiths Market*], 1974. Tekin Yayınevi, 1980.

———. *Akçasazın Ağaları 2, Yusufçuk Yusuf,* 1975. Yapı Kredi Yayınları, 2004.

———. *Bir Bulut Kaynıyor: Bu Diyar Baştanbaşa 4.* Yapı Kredi Yayınları, 2003.

———. "Yaşar Kemal Anadolu Doğasını Yeşil Atlas'a Anlatmıştı." Reportage, Güven Eken and Güneşin Aydemir, 2003. https://www.atlasdergisi.com/kesfet/kultur/yasar-kemal-anadolu-dogasini-yesilatlasa-2003-anlatmisti.html

———. *Yaşar Kemal on His Life and Art.* Introd. and Notes Barry Tharaud. Syracuse University Press, 1999.

Zedler, Joy B. "The Continuing Challenge of Restoration." *The Essential Aldo Leopold: Quotations and Commentaries,* edited by. Curt Meine and Richard L. Knight. University of Wisconsin Press, 1999, pp. 116–126.

Chapter 13

Affective Ecocriticism

From Thinking to Feeling and Being Together

Denis Petrina

NO MORE REASON FOR REASON, OR WHY AFFECT?

The emergence of the category of affect in the discourse of critical eco studies marks a deliberate and conclusive break from the long-living Cartesian tradition. The Cartesian project, whose essence could be succinctly put into a formulated by Descartes formula "*cogito ergo sum*" (I think therefore I am) (28), has brought the concept of the rational, autonomous, self-sufficient—thus, epistemologically and ontologically superior—*human subject* to the forefront. However, affect theory has highlighted the problematic nature of the preoccupation with anthropocentric rationality. In short, the anthropocentric "delusion of grandeur" (deeply rooted in the false belief of the superiority of human reason) has led to a global environmental catastrophe, leading to accelerated degradation of ecosystems and the gradual destruction of the conditions of life on Earth: not *despite* reason, but even *because of* it. Human reason (believed to be the tool of emancipation) has become the hostage of exploitative capitalist ethics, legitimized by scientific instrumentalization and objectification of nature. What is more, the Cartesian project has proved to be utterly incapable of generating proactive and transformative ethical, social, and political scenarios—and, in this sense, only deferring the threat of what now might be conceived as the inevitable.

What happens when we move from *reason* to *affect*? Why affect? And what, after all, is affect? First and foremost, affect does not equal emotion: as argued by Brian Massumi, emotion is only a subjectified and rationalized remainder of affective content (28)—thus affect is not something that is uniquely human. To experience affect, one does not need to be a human, to have reason, or even, strictly speaking, be "alive"—instead, what one needs is to be *entangled* in a network of relations with *others*. This entanglement, relationality, co-, inter-, infra-, . . . dependence is the starting point of what I call in this paper "affective ecocriticism." The introduction of the affective dimension into ecocriticism allows us to rethink our "locatedness" in larger ecosystems from the perspectives of interconnectedness (*immanent ontology,* Spinoza), molecular actors (*"becoming-molecule,"* Deleuze and Guattari), caused by the feeling of "doom and gloom" negative sentiments (*"affect aliens and killjoys,"* Ahmed), speculative analytics (*"staying with the trouble,"* Haraway) and, ultimately, based on those accounts, to attempt to answer the question of how we could imagine a brighter tomorrow (*ethics of care,* Grosz, Braidotti).

This paper is structured around the dynamic from the (illusionary) *ego* to the (radical) *eco*: first, I am to sketch a conceptual model that was developed by Spinoza and is centered around affect; then, I will examine a repertoire of negative affective responses to the environmental crisis: from passive and apathetic material and emotional depletion to more "militant" "environmentalist killjoy"; finally, I will explore how "staying with the trouble" becomes a promise of a new onto-epistemo-ethical regime whose focal point is the acknowledgment of shared vulnerability and thus the necessity of care.

THINKING "ECOLOGY OF BELONGING" WITH SPINOZA: AFFECTIVE ONTOLOGY

Baruch (Benedictus de) Spinoza's immanent ontology appears as a perfect contender to explain the complex interrelations between affect, being, and ecological thought. The essence of Spinoza's philosophy consists in his famous postulate *Deus sive natura* (God or nature, God as nature), which negates the idea of the outside (a "transcendental plane") and, instead, emphasizes the concept of *divine* (metaphysical) nature expressed through the two-fold dynamic of *natura naturans* (nature as an active force) and *natura naturata* (nature as a passive product of itself). In other words, nature is both the cause and the result of itself, which means that it is *immanent.*

What inhabits the plane of immanence—the plane of self-producing and produced by self-nature—is its *modi*—bodies. Bodies are the variations of the singular substance of nature and should thus be perceived as the expressions of the substance: bodies could be organic and inorganic, living and non-living, moving and resting; people, animals, trees, rocks, waves, stars, molecules—everything is a body in Spinoza's ontological system. These individuals/bodies, however, are never separate: the idea of immanence presupposes mutual reciprocity of the bodies that together constitute the plane of immanence.

The *token* or *unit* of this reciprocity is affect, which, as defined by Spinoza in his treatise titled *The Ethics*, is "the modification of the body, whereby the active power of the said body is increased or diminished, aided or constrained, and also the ideas of such modifications" (Ch. 3, Def. III) Let us stop here to articulate the key features of affect. First and foremost, affect is greater than an emotion: while the latter is "the idea of the modification," the former is the modification of the body that includes the idea as the imprint. Second, since affect is defined as a "modification," it is never purely subjective, personal, and static, but instead *intersubjective, relational,* and *dynamic*: bodies interact with each other and this mutual interaction causes them to affect and be affected by each other—thus, modify each other. Third, the modifications bodies undergo are never neutral: as Deleuze suggests, affects-modifications (what he, after Spinoza, calls *affecto*) enable affects-transitions (*affectus*) (*Spinoza*, 26–27). The latter let the body transition into a greater or a lesser degree of perfection: in the first case, the body's capacities to act increases, in the second—diminishes.

This means that affect is the derivative of power. Here, however, power should not be construed as crude force, domination, selfish appropriation of resources, or an intention to exploit, manipulate or use others for one's own advantage. For Spinoza, everybody possesses a certain degree of power due to the fact that everybody can affect other bodies. Since Spinozian ontology is relational ontology *par excellence*, the utmost degree of power is following the divine/natural order: being more active, increasing one's capacities to act and in so doing increasing others' capacities to act. Figuratively speaking, for Spinoza, there is no (and should not be) singular Power with a capital P, only powers: negative *power over* (*potestas*) should be superseded with *power to* (*potentia*). As noted by Andrea Zevnik, the very being in Spinozian thought originates from power (*potentia*) (60), which means that the correct conceptual framework for understanding Spinoza is not the one that centers around subjugation, exploitation, and destruction, but that of empowerment, facilitation, and creation—in other words, an *affirmative* celebration of being. Because of that high degree of responsibility (to both ourselves and others), Rosi Braidotti reasonably calls Spinozian model of the world the "ecology of belonging" (*Nomadic Ethics,* 174).

To better articulate this ecological aspect of Spinoza's thought, we should translate Spinoza's ontological observations into an ethical-political scheme. Spinoza's focus shifts from our personal accomplishments and ability to wield power (*potestas*) to a matter of composition: how we can utilize our power to enter better compositions (both with ourselves and others—both human and non-human) and, vice versa, how we can form compositions that help us and others utilize our powers better. This communal aspect of Spinozian thought is what makes it truly ecological. Here, affect is the category that removes the human subject from the pedestal of being: if affectivity is a property of any body, then it decenters the human subject since it has no apparent privilege among other bodies. This, in its turn, results in a model of flat ontology, where *ego* is superseded with *eco*, subjectivity with relationality, and rigid pragmatics with exploration and situational and variative ethics.

Spinoza, read through the lens of ecocriticism, reminds us not only about the illusionary nature of human Power, but also about the dangers and potential repercussions of identifying with the illusions of this seemingly uncontrolled Power. For Power brings nothing but sad passions, it poisons both the individual and the world by fostering passivity, apathy, conformism; in the context of contemporary Capitalocene, it fosters greedy exploitation of natural resources, cynical indifference to pressing ecological problems, and mindless destruction of living environments. Eccy de Jonge explains: "Spinoza's politics aims to protect those who would seek to live in harmony with others against the tyranny of those whose private concerns and opinions take precedence" (130).

Thus, the bottom line of Spinoza's philosophical teachings is freedom, which can transform sad passions into active, enhancing our capacities to act affects. Freedom, as Elisabeth Grosz perfectly puts it, "is the freedom to act, encounter, enhance, learn, grow, make do, with as many of those objects, material and conceptual, that agree with our natures as we are able" (91). Freedom, therefore, is a space of constant learning, exploration, experimentation, formation of various compositions whose longevity and efficiency is tested by trial and error; the only applicable rule is that of the divine/natural order: we only truly empower ourselves if we contribute to the empowerment of others.

ENJOYING "CRUEL OPTIMISM" VS. KILLING JOY: NEGATIVE AFFECTS AND ALIENS

Viewed through the Spinozian lens, affect is ecological since it shifts our focus from the universal(izing) human perspective to, as we could call, following Deleuze and Guattari, the molecular: barely visible yet tangible interrelations, interconnections and intersections, undetermined situatedness, intensive becomings (with others), dynamic compositions and decompositions that constitute being in the world (275). Situated in the ecological context, affect encourages us to rethink our belonging to the world through such categories as three—environmental, mental, and social—ecologies (*Three*

Ecologies, 68–69), assemblages we find ourselves in and form with others (8), collective and personal moods, permanent and fleeting feelings, inspiring and suffocating atmospheres. However, the common affective vector (which, figuratively speaking, is only a barometer and only indicates the atmosphere that encompasses a broad spectrum of multiple affective responses), is *negative*. The ecological calamities, crises, catastrophe "seem too big to process emotionally," as Kyle Bladow and Jennifer Ladino rightfully point out (16).

The refusal to accept and act upon the omnipresent ecological issues is an instance of what Lauren Berlant calls "cruel optimism," which, too, is within the spectrum of affective responses. As Berlant explains it, what makes this optimism cruel is people's refusal "to interfere with varieties of immiseration" and their (un)conscious choice "to ride the wave of the system of attachment they are used to" (23). It is blissful yet potentially dangerous ignorance which leads to nothing but the exacerbation of the detrimental situation. Locating the conundrum of cruel optimism within the Spinoza's scheme, the naïve belief in Power—what governments and corporations tell us about the 'magical remedy' of conscious production and consumption and similar panacea-like scenarios is the pitfall, in which we have been trapped for too long.

I am, therefore, more interested in affects that emerge beyond the spectacle of the Capitalocene—affects that are essentially positioned *against* it. The spectrum of these *critical* affects spans between two seemingly different yet interrelated poles: the passive, which I place under the umbrella term of "depletion"—apathy, melancholia, despair, hopelessness and helplessness; and the active, which is characterized by the attempt to utilize the negative sentiments and weaponize them against deeply-rooted systematic neglect. Both these affective constellations point to the problem of alienation: the vicious dominant ideology of the "world-for-us" (humans) has led to the predicament in which the world has turned "against us"; there is, thus, the need to rethink (through our affective encounters and reactions) the very concept of the "world-for-us" and expand the notion of "us" so that it could incorporate not only human ecosystems, but the whole biosphere.

The concept of depletion is of particular methodological value, as it emphasizes both the material (that of natural resources) and emotional (feeling hopeless and helpless) aspects of depletion. Graig Uhlin articulates this solastalgic nexus the following way: "Ecological collapse prefigures emotional collapse. Environmental crisis registers as an atmospheric shift where a restorative or life-sustaining atmosphere turns suffocating and draining" (282). In Berlant's words, this affective configuration could be called 'flat affect' when intrinsic to affect vitality and intensity evaporate and what is left is only an "empty shell" (qtd. in Duschinsky and Wilson, 185). The paradox, however, is that flat affect occurs as a response to too high intensity, due to which the subject experiences overbearing shock and thus seeks to protect itself by withdrawing into itself. Put differently, emotional depletion/flat affect functions as a defense mechanism against the traumatic experience of the real. It is, indeed, this traumatic experience that causes the subject's resignation and melancholia; melancholia, however, is different from indifference that always accompanies cruel optimism: unlike the indifferent (blissfully ignorant) subject who simply *does not* care, the melancholic subject cares *too much*. Again, as Uhlin notes, such a response "to environmental harm can function as diagnostic . . . of our situation" (295), in other words: as a necessity to operationalize care to overcome the paralyzing feeling of 'doom and gloom' and move from *diagnosis* to *praxis*.

Praxis starts with the proposed by Sara Ahmed strategy of alienation: the refusal "to share an orientation toward certain things as being good" (39). She introduces the concept of "affect aliens": "feminist killjoys, unhappy queers, and melancholic migrants" (30) in order to conceptualize this different orientation toward what is considered to be "the common good." At first sight, it might seem that affect aliens become such against their own will: they are *made* aliens because they are radical others. Be that as it may, another (more productive) dimension of alienation is a deliberate attempt to oppose the *status quo*, in which affect plays a key role: it inevitably highlights the discrepancy between *what is lived* and *what is felt*, and in this regard, acts as an indicator of the unresolved tension between the two. This dissonance, therefore, serves as a catalyst for change: not for "common good," which is hegemonic and masks the interests of particular individuals, but at least for the better.

Ahmed demonstrates that being "alien" (especially in the context of "cruel optimism") is tantamount to being a "killjoy." Killing joy appears to be one of the few valid strategies to undermine the "smoke and mirrors" spectacle of the Capitalocene. Not only do alienation and killing joy distance us from the commonly wrought environmental havoc, they also place a heavy burden of showing others what is wrong (and therefore teaching others what should be done about it). The provided by Ahmed list of affect aliens could be supplemented with a few other significant for environmentalism figures: environmentalist killjoys (Bladow and Ladino, 11), tireless doomsayers, idealistic activists. And yet, the problem with activism is that it has taken the form of guerrilla activity. While without any doubt eco-activism makes crucial and so painfully needed changes politically, without the revision of ethical and epistemological premises of the crisis, it remains a battle against fleeting time.

VULNERABILITY, SUSTAINABILITY, EMPOWERMENT: ON THE ETHICS OF AFFECT

While actions must, indeed, be taken, there is a preceding step that almost always remains ignored: *thinking* before *acting*. Even though this warning might look overly simplistic at first glance, it, in fact, emphasizes the transformative power of thinking: not thinking as we are used to (such a modality of thinking proved itself of little help, as its best outcome is the alleviation of the damage already done), but thinking radically *otherwise*. What would, then, be a radically different ecological thought? Paradoxically, a transition to ecological thought would require the renouncement of our rational agency and a conceptual turn to affect. Affect allows us to perform two crucial conceptual operations, which naturally stem from the reconfiguration of thought. Ecological ethics is based upon a paradoxical quasi-dialectical model "no subject/everybody is a subject." The first scenario lets us overcome *egology*—the Cartesian mode of thinking that reinforces the privilege of the human subject. The second scenario treats everybody (every modus, to quote Spinoza) as a potential actor and thus reveals the agentic nature of the world around us. The first scenario is an exercise in humility, reminding us of our limitations, the second is an exercise in responsibility, teaching us to seek harmony in our connections with other actors.

However, we should not be mistaken that affect always translates into *power to*. We find ourselves in an intricate network of connections, which can both enhance our capacities to act as well as empower others and diminish them, make us weak, paralyze and even harm or destroy us. To understand affect as necessarily affirmative would be to only get a half of the story right; instead, what affect points to is our *common vulnerability*. In the existential (or even ontological) situation of affective entanglement, this shared vulnerability translates into, as Braidotti perfectly put it, "being-there and being-in-relation to others" (*Nomadic Ethics,* 174). As she reminds us: only by fostering the connections, relations, compositions, and kins based on mutual responsibility, solidarity, and care, can we resist the disastrous expansion of biopolitics and necropolitics, which have been destroying our living world(s) (*Posthuman,* 118). Only by acknowledging our and others' beings and becomings can we activate the untapped potential of ecological thought: "new concepts, affects, and planetary subject formations" (104). As beautifully worded by Sara Ahmed, vulnerability reminds us of our existential co-dependence, which implies two potential outcomes: either loss or pain, or love and care (*Politics of Emotions,* 68–69, 125).

The outcome of our encounters with others can never be forecasted or determined. Neither pain (luckily) nor joy (sadly) is a given. Affect means openness, and openness, in its turn, means a broad range of potential scenarios, each of which essentially transforms us and others. Thus, how to sustain our being while undergoing transformations, modifications, mutations? Braidotti offers an existential-ethical formula: "affectivity is the propensity for changes and transformation that is directly proportional to the subject's ability to sustain the shifts without cracking" (*Posthuman Affirmative Politics*, 175). The essence of this ethical lesson is to incorporate breaks, protect ourselves (as well as others) while transitioning from one state to another, not only merely *sustaining* every shift, but also enhancing our (and others) perseverance, resilience, and endurance—thus *common* capacity to act as the imperative of being together.

IN LIEU OF CONCLUSION: CARE FOR THE LIVING, CARE FOR THE DEAD

Because of the urgent need for a change, our ethical obligation is, to borrow Donna Haraway's phrasing, *to stay with the trouble*. "Staying with the trouble requires making oddkin; that is, we require each other in unexpected collaborations and combinations, in hot compost piles" (4). Our rational agency is of little help in staying with the trouble; in fact, it would not be an overstatement to say that our rational agency is partly responsible for getting us into this trouble. The turn to affect—as something we all share—would, thus, mean shifting away from consciousness to sensibility, from the category of seemingly self-sufficient human subject to that of the open and vulnerable body, from imposed from above measures to horizontal mutual learning and experimentation. There is an integral aesthetical component of this ecological ethical paradigm, which Guattari spoke of (*Chaosmosis*, 107) aesthetic directly answers this call of experimentality, as it encompasses sensibility, feeling, and empathy—empathy not as feeling for, but as feeling with (Bladow and Ladino, 13). What we, therefore, need is, according to Braidotti, the restoration of the lost feeling of "intimacy with the world" (*Posthuman Affirmative Politics*, 51).

To restore intimacy with the world we need to rethink the premises of our being in the world, being together. Jacques Derrida's conceptualization of being as *l'ontologie* (ontology) and *l'hauntologie* (hauntology) could be utilized as an important reminder of the absence of *givenness* and *presence*: the seeming paradox is that nothing is given and we never *are*, at least in a way we think of our being—being self-sufficient, independent, and fully aware of the world and the impact we make on the world (and *vice versa:* the impact the world makes on us). Invisible spectral threads connect us to others—those who are not necessarily human, nor animals, plants, bacteria, or even molecules. Even more so, they are not necessarily living or even alive. For Derrida, the true meaning of responsibility is being responsible not only for the present (what is now), but also for the future (the things to come), as well as for the past (what has been done) (*Specters*, xviii); being *truly* responsible entails being responsible for specters and ghosts—for those who technically *are not*. In a Derridean manner, we could say that this responsibility *haunts* us. We must face it and be responsible not only for the world as it is now, but also for the unrepairable ecological damage, for inevitable repercussions, for all the species: both on the verge of extinction and already extinct. Because it is not a future obligation anymore: it is a debt that has long been overdue.

REFERENCES

Ahmed, Sara. *Happy Objects*. Edited by Gregg, Melissa, and Seigworth, Gregory. *The Affect Theory Reader*. Duke University Press, 2010, pp. 29–51.

———. *The Cultural Politics of Emotions*. Edinburgh University Press, 2004.

Anderson, Ben. "Affective Atmospheres." *Emotion, Space and Society,* vol. 2, no. 2, 2009, pp. 77–81.

Berlant, Lauren. "Cruel Optimism." *A Journal of Feminist Cultural Studies*, vol. 17, no. 3, 2006, pp. 20–36.

Bladow, Kyle, and Ladino, Jennifer. *Toward an Affective Ecocriticism: Placing Feeling in the Anthropocene.* Edited by Kyle Bladow and Jennifer Ladino. *Affective Ecocriticism: Emotion, Embodiment, Environment.* University of Nebraska Press, 2021, pp. 1–24.

Braidotti, Rosi. *Nomadic Ethics*. Edited by Daniel W. Smith and Henry Somers-Hall. *The Cambridge Companion to Deleuze*. Cambridge University Press, 2012, pp. 170–197.

———. *Posthuman Affirmative Politics.* Edited by Stephen Elliot Wilmer and Audronė Žukauskaitė. *Resisting Biopolitics: Philosophical, Political, and Performative Strategies*. Routledge, 2016, pp. 30–56.

———. *The Posthuman*. Polity Press, 2013.

Deleuze, Gilles. *Spinoza: Practical Philosophy*. Translated by Robert Hurley. City Lights Book, 1988.

Deleuze, Gilles, and Guattari, Félix. *A Thousand Plateaus: Capitalism and Schizophrenia*. Translated by Brian Massumi. University of Minnesota Press, 2004.

Derrida Jacques. *Specters of Marx: The State of the Debt, the Work of Mourning and the New International*. Translated by Peggy Kamuf. Routledge, 1994.

Descartes, Rene. *Meditations on First Philosophy*. Translated by Elizabeth Haldane. *The Philosophical Works of Descartes*. Cambridge University Press, 1996.

Duschinsky, Robbie, and Wilson, Emma. "Flat Affect, Joyful Politics, and Enthralled Attachments: Engaging with the Work of Lauren Berlant." *International Journal of Politics, Culture, and Society*, vol. 28, no. 3, 2014, pp. 179–190.

Grosz, Elizabeth. *The Incorporeal: Ontology, Ethics, and the Limits of Materialism*. Columbia University Press, 2017.

Guattari, Félix. *Chaosmosis: An Ethico-Aesthetic Paradigm.* Translated by Paul Bains and Julian Pefanis. Indiana University Press, 1995.

———. *Three Ecologies*. Translated by Ian Pindar and Paul Sutton. The Athlone Press, 2000.

Haraway, Donna. *Staying with the Trouble: Making Kin in the Chthulucene*. Duke University Press, 2016.

Jonge, Eccy de. *Spinoza and Deep Ecology: Challenging Traditional Approaches to Environmentalism.* Routledge, 2004.

Massumi, Brian. *Parables for the Virtual: Movement, Affect, Sensation.* Duke University Press, 2002.

Spinoza, Benedict. *The Ethics*. Translated by Elwes R. H. M. Blackmask Online, 2001.

Thrift, Nigel. *Non-Representational Theory: Space, Politics, Affect*. Routledge, 2008.

Uhlin, Graig. *Feeling Depleted: Ecocinema and the Atmospherics of Affect*. Edited by Kyle Bladow and Jennifer Ladino. *Affective Ecocriticism: Emotion, Embodiment, Environment.* University of Nebraska Press, 2021, pp. 279–298.

Zevnik, Andreja. *Lacan, Deleuze, and World Politics: Rethinking the Ontology of the Political Subject*. Routledge, 2016.

Chapter 14

Empirical Ecocriticism

Unveiling the Power of Literature in Environmental Consciousness

Milena Škobo

Empirical ecocriticism, as an ecocritical theory, holds a significant place in the broader field of ecocriticism, offering a fresh perspective on the intricate relationship between literature and the environment. Its emergence represents a critical response to the limitations and criticisms of traditional ecocritical approaches, which often relied on subjective interpretations and lacked empirical validation. In contrast, empirical ecocriticism adopts "an empirically grounded, interdisciplinary approach to environmental narrative" (Schneider-Mayerson et al., "Empirical Ecocriticism" 328), bridging the gap between ecocriticism and empirical research.

The significance of empirical ecocriticism within the broader field lies in its three fundamental features. Firstly, it employs a rigorous and verifiable approach to studying the ecological dimensions of literature, relying on both quantitative and qualitative data analyses to enable a more concrete and evidence-based analysis and interpretation of environmental themes in literary works. Secondly, this reliance enhances the validity and credibility of ecocritical interpretations, introducing objectivity to the analysis and fostering a robust scholarly dialogue. Lastly, empirical ecocriticism's ability to foster interdisciplinary collaboration by integrating perspectives and methodologies from diverse disciplines such as environmental studies, geography, and cognitive science is paramount. For instance, in exploring the impact of climate fiction on readers' attitudes and behavior or investigating how narrative empathy influences concern for nonhuman species, empirical ecocritics employ original research methods like interviews, focus groups, surveys, or controlled experiments. Given the expertise required in both ecocriticism and social scientific methods, conducting empirical studies in ecocriticism often involves collaboration among researchers, enriching the field by combining insights from different domains and providing a comprehensive understanding of the ecological implications present in literary works.

The emergence of empirical ecocriticism as an ecocritical theory has taken the understanding of the power of literature in shaping environmental consciousness to a whole new level, offering a unique and evidence-based perspective on the transformative impact of environmental narratives. Unlike traditional ecocritical analysis, empirical ecocriticism extends its scope to encompass various narrative media, including novels, short stories, poetry, children's literature, film, television, video games, music, and theatre. Central to this analysis are the formal and aesthetic elements within texts, such as narrative voice, perspective, genre, fictionality, and character construction, which influence readers' ecological consciousness.

The dynamic and multifaceted nature of this ecocritical theory is exemplified by case studies conducted by pioneering researchers in this field. The majority of research has leaned towards quantitative approaches (Małecki et al., "Literary Fiction"; Małecki et al., "Feeling for Textual Animals"), while a smaller number of studies have incorporated qualitative data through open-ended survey questions (Monani et al., "Loving Glacier"; Schneider-Mayerson, "The Influence of Climate Fiction"; Schneider-Mayerson, "Just as in the Book?") or focus group interviews with questionnaires (Breraton and Gómez, "Media Students"). However, the emphasis on quantifiable scientific evidence has led to concerns about potential reductionism and a lack of theoretical depth from scholars adhering to hermeneutics. To avoid marginalisation within the humanities and maintain conceptual and theoretical rigor, finding a balance between empirical methodologies and hermeneutic approaches becomes essential. In this context, qualitative approaches, such as phenomenology, ethnography, and grounded theory, have been identified as promising avenues within empirical ecocriticism (Sopcak and Sopcak, "Qualitative Approaches"). These qualitative methods serve as a bridge, providing solid conceptual and theoretical foundations for the emerging field of empirical ecocriticism, while also acknowledging the importance of empirical evidence in exploring the complex relationship between literature and environmental consciousness.

In the pursuit of illuminating the potential contributions of empirical ecocriticism to ecocriticism and environmental humanities, notable studies employing both quantitative and qualitative approaches serve as exemplars. One such illustration is found in the article titled "Empirical Ecocriticism: Environmental Texts and Empirical Methods" by Schneider-Mayerson and colleagues, where three pilot studies are showcased (3–7). These studies effectively demonstrate the seamless integration of empirical research methods, such as questionnaires, focus groups, and experiments, with the analysis of environmental texts. Through this integration, empirical ecocriticism adeptly investigates the profound impact of environmental texts on audiences, thereby demonstrating the power and efficacy of such methodologies in this emerging field.

The first case study conducted by Matthew Schneider-Mayerson focuses on exploring the influence of climate fiction on American readers' awareness of climate injustice and perceptions of climate migrants. The aim of the study was to evaluate specific claims made about climate fiction by both ecocritics and authors. His methodology centres on utilizing detailed questionnaires, which is similar to his previous studies on eco-apocalyptic disaster movies (Schneider-Mayerson, "Disaster Movies") and climate fiction (Schneider-Mayerson, "The Influence of Climate Fiction"). Through administering a questionnaire to readers of Paolo Bacigalupi's dystopian cli-fi novel *The Water Knife* (2015), valuable information about the readers' responses was gathered, shedding light on the influence of factors such as age, ethnicity, and gender on their identification and empathy with characters. The study revealed that the novel increased awareness of climate injustice and its human consequences. However, it also unveiled a noteworthy finding—some liberal readers ended up fearing climate migrants rather than empathizing with them (Schneider-Mayerson et al., "Empirical Ecocriticism" 329).

The second case study is conducted by Małecki, Weik von Mossner, and Dobrowolska, who adopted a quantitative experimental methodology, the approach initiated by Małecki through his research on the attitudinal and behavioral impact of animal narratives in literature (see Małecki, "Experimental Ecocriticism"; Małecki et al., "Literary Fiction"). The authors explored why Alice Walker's story "Am I Blue?" in particular displayed contrasting effects compared to other animal narratives studied. They proposed a compelling hypothesis, suggesting that the lack of impact on attitudes toward animals in general could be linked to Walker's approach of drawing parallels between the oppression of underprivileged human groups and the oppression of nonhuman species in her text. This hypothesis was explored by employing three kinds of evidence: First, they analysed the narrative structure of Walker's essay, drawing attention to its strategic use of trans-species empathy and the various parallels between human and animal suffering. Second, they examined the controversy surrounding the 1994 decision of the California State Board of Education to ban the essay from a state-wide test for tenth graders. Lastly, they compared how "Am I Blue?" affected readers' views on animals with its impact on their views on human minorities.

The third case study referred to in the article "Empirical Ecocriticism" by Schneider-Mayerson and co-authors is the one conducted by Pat Brereton and Maria Victoria Gómez. The authors expanded the traditional focus of ecomedia on film and television to explore the reception and impact of shareable videos popular with younger generations. They employed a methodology that combined focus group interviews with questionnaires, allowing them to gather valuable insights from Third Level students in Ireland. Through an expert reading of Prince Ea's "Dear Future Generations: Sorry," a climate change-themed spoken word music video with over 23 million views, they discovered that most students preferred direct-address speeches and persuasive documentary videos.

Through the first pilot study, we witness how empirical ecocriticism serves as a powerful tool in gathering valuable insights about readers' responses to climate fiction. The study reveals that readers' identities play a pivotal role in shaping their empathetic connection with characters, shedding light on the diverse perspectives that emerge when encountering environmental literature. In the second pilot study, we gain a deeper understanding of the complexities surrounding literature's ability to foster empathy and environmental consciousness. The nuanced finding of this research reminds us that literary works possess intricate and context-dependent effects on environmental awareness, and their influence cannot be generalised across all themes and scenarios. The third pilot study demonstrates the versatility of empirical ecocriticism, incorporating qualitative research methodologies to understand media consumption patterns. This innovative approach reveals the potential for empirical ecocriticism to adapt to new and emerging media platforms, ensuring its relevance in an ever-changing media landscape. These studies highlight the fact that empirical ecocriticism is a powerful force in illuminating the impact of environmental literature on readers' attitudes, behaviour, and consciousness.

Another pioneering study in empirical ecocriticism sheds light on the rapid development and maturation of the discipline of ecocriticism in the early twenty-first century and reinforces the notion that ecocriticism was undergoing significant evolution. The forthcoming article by Scott Slovic and David Markowitz titled "Tracing the Language of Ecocriticism: Insights from an Automated Text Analysis of ISLE: Interdisciplinary Studies in Literature and Environment," offers insights into the language of ecocriticism through the Linguistic Inquiry and Word Count (LIWC) method for automated text analysis of the journal *ISLE* from 2004 to 2018 (Lifu and Slovic 86). The study reveals a trend towards increased abstraction and the use of jargon, reflecting the discipline's growing sophistication and the creation of a specialized theoretical vocabulary. By utilizing automated text analysis, the researchers provide empirical evidence of language trends within ecocritical discourse, contributing to a deeper understanding of the field's growth and intellectual trajectory. The article's findings and methodology align with the principles of empirical ecocriticism by employing quantitative analysis to explore patterns and trends in the field. The integration of digital humanities techniques and automated text analysis highlights the interdisciplinary nature of empirical ecocriticism and its potential to uncover valuable insights through empirical research strategies.

The potential for further research in empirical ecocriticism is extensive, particularly in examining audience responses to specific types of texts. Notably, the research papers by Škobo and Đukić on J. G. Ballard's urban violence novels provide valuable insights and pave the way for future research in the field, as they offer an opportunity to explore how literature portraying alternative ecological futures can influence readers' perceptions of urban landscapes and their implications for environmental and public health. Stacy Alaimo's concept of the environment as human landscape "in which people blend into their living spaces" ("Bodily Natures" 122) finds resonance in these novels, reflecting the violence and environmental illness that pervade hi-tech zero-waste communities, resulting in spiritual, emotional, and mental paralysis among their inhabitants (see Škobo and Đukić, "James G. Ballard's Urban Violence"; Škobo and Đukić, "Manifestations of 'New-Age' Religions"; Škobo and Đukić, "Urban Landscape"). Understanding the intricate connection between literature, empirical ecocriticism, and reader behaviour, encompassing attitudes towards urban environments and identification of psycho-social risks threatening mental health, becomes essential in revealing the potential transformative power of literature. The focus on narratives depicting urban violence offers fresh perspectives on urban environments and their impact on public mental health. Unveiling how readers perceive and engage with these narratives may open pathways for cultivating more sustainable and ecologically conscious attitudes and behaviours within urban communities.

Regarding disability studies and environmental illness (EI) or Multiple Chemical Sensitivity (MCS), empirical ecocriticism holds significant promise in providing valuable insights and solid evidence on the complex relationship between human bodies, the environment, and the impacts of harmful substances. Through empirical research methods, empirical ecocritics can examine how certain literary works portray or engage with the effects of built environments and harmful substances on human health and disability. They can investigate how literature reflects and responds to the real-world experiences of individuals living with EI or MCS, as well as the challenges they encounter while coexisting with potentially hazardous material agencies.

In a similar vein, another prominent direction in empirical ecocriticism is the investigation of the impact of pandemic literature on readers' perceptions of precarity and vigilance concerning public health and ecological threats. Scott Slovic's work serves as a prime example of this research direction, as he focuses on developing empirical ecocritical studies using pandemic texts to instil a sense of "healthy precarity" in readers (Lifu and Slovic 85). In an interview, Slovic discusses his ongoing research and aims to understand whether readers of pandemic novels, exposed to extreme human vulnerability, exhibit greater caution and care in other aspects of their lives. By exploring pandemic texts, Slovic believes that readers can develop a heightened sense of vulnerability that may lead to more cautious and mindful behaviour, not only in the context of disease but also in relation to global climate change (85). The intersection of empirical ecocriticism and pandemic literature research exemplifies the power of this eco-concept in illuminating how literary works can drive transformative change, motivating readers towards adopting sustainable behaviours and practices.

Expanding Scott Slovic's research direction on the impact of pandemic literature on readers' environmental consciousness, empirical ecocriticism can focus on conducting cross-cultural analysis to explore how readers from diverse cultural backgrounds respond to these narratives. By examining pandemic literature from different countries and regions, such as *Rabies* (1983) by Borislav Pekić in Serbia, *Blindness* (1997) by José Saramago in Portugal, or *The Plague* (1947) by Albert Camus in France, alongside contemporary pandemic books from 2020 and beyond, researchers can conduct parallel experiments or longitudinal studies to understand the reasons behind delayed reactions to environmental themes and the evolving impact of pandemic literature on readers' environmental consciousness. Cross-cultural comparisons of environmental narratives also offer intriguing opportunities, uncovering the impact of culturally specific factors on eco-consciousness across societies.

The future of empirical ecocriticism aligns with the latest developments in the field of ecocriticism, echoing Scott Slovic's call to "go public" (Lifu and Slovic 92–93). Inspired by Marybeth Gasman's 2016 book *Academics Going Public: How to Write and Speak Beyond Academe*, Slovic's adoption of this concept urges ecocritics to communicate their research to broader non-academic audiences, breaking away from the conventional focus on specialised scholarly publications. This emphasis on *going public* harmonizes with the transformative potential of empirical ecocriticism, extending its impact beyond academia and encouraging greater societal engagement with environmental themes. By adopting accessible writing styles and engaging with non-academic platforms such as op-eds and press releases, as suggested by Slovic in discussing the latest developments in the broader field of ecocriticism, empirical ecocritics can disseminate their findings, contributing to broader environmental conversations.

As empirical ecocriticism looks ahead, it embraces cutting-edge technologies and innovative methodologies, offering deeper insights into literature's profound impact on ecological consciousness. This integration of empirical research methods requires meticulous consideration of data selection, interpretation, biases, and ethical aspects. It is essential to avoid reducing literary works to mere data points and to appreciate the nuanced and subjective dimensions of artistic expression. Balancing the critical rigor of empirical analysis with the artistic essence of literature (with rich literary and cultural contexts) necessitates constant reflection and refinement.

Moreover, as empirical ecocriticism continues to evolve, it will increasingly prioritise global and postcolonial perspectives, recognizing the importance of diverse voices in exploring ecological themes within literature. This approach opens new research opportunities and highlights how environmental issues intersect with social, cultural, and historical contexts. Through the application of social scientific methods, empirical ecocriticism uncovers the social, psychological, and political impacts of environmentally engaged narratives on audiences, catalyzing transformative changes in attitudes, behaviour, culture, infrastructure, and policy—a crucial response to the pressing challenges of climate change and mass extinction.

In addition to these priorities, empirical ecocriticism delves into exploring the potential of immersive virtual reality experiences and data-driven storytelling, incorporating real-time environmental data to amplify emotional resonance and effectiveness, forging deeper connections between audiences and ecological concerns. Also, the integration of artificial intelligence and machine learning techniques in textual analysis holds promise in uncovering previously overlooked ecological themes in literature.

One primary concern regarding empirical ecocriticism is the risk of reductionism. Its reliance on quantitative data and analysis, while aiming at objectivity, can oversimplify the nuanced and multifaceted aspects inherent in both literary and environmental issues. Despite the aspiration for objectivity, subjectivity can still influence data interpretation, introducing potential bias into the analysis. This strong emphasis on empirical data may also run the risk of marginalizing the aesthetic and artistic qualities that are intrinsic to literature, reducing literary works to mere repositories of data. Furthermore, empirical ecocriticism often prioritizes data that is easily quantifiable, potentially overlooking the richness that can be derived from qualitative analysis and close readings of texts. The bias towards quantitative methods may lead to the underrepresentation of qualitative and mixed-method approaches, thereby limiting the scope of inquiry. Therefore, it is crucial to consciously strive for a balanced integration of both quantitative and qualitative methods. While quantitative data is valuable, qualitative analysis and close readings of texts can provide depth and nuance to the interpretations, ensuring a more comprehensive understanding of literary and environmental issues.

The interdisciplinary nature of empirical ecocriticism poses its own set of challenges. Researchers are often tasked with bridging the gap between their expertise in literature and their proficiency in empirical research methods, a process that can be resource-intensive and time-consuming. Therefore, it is essential to avoid a bias towards quantitative approaches and recognize that qualitative and mixed-method approaches can offer distinctive insights.

Ensuring ethical considerations in data collection and obtaining participant consent must be a priority, especially when dealing with sensitive environmental and societal topics within empirical ecocritical studies. When conducting a study examining the impact of industrial pollution on local communities, researchers would need to prioritize ethical considerations by ensuring that the data collection process respects the privacy and well-being of the affected individuals. Obtaining informed consent might involve transparent communication about the study's goals, potential risks, and the rights of participants. This ethical approach ensures that the research contributes valuable insights without causing harm to the individuals or communities involved.

Moreover, the pursuit of generalizable findings presents a challenge given the nuanced, context-specific nature of literature. Literature, serving as a reflection of diverse human experiences and cultural subtleties, often resists easy categorization into universally applicable conclusions. For instance, researchers exploring the representation of climate change in literature across different cultures is a challenging task. The challenge arises in the nuanced nature of literary works, where cultural, historical, and linguistic factors play significant roles. One cannot easily generalize findings across diverse contexts, as a metaphorical portrayal of environmental issues in one cultural setting might differ significantly from another. Researchers must acknowledge and navigate this challenge, adopting a more context-sensitive approach in their analysis, recognizing that conclusions may vary based on the unique characteristics of each literary and cultural context.

Additionally, the analysis of historical texts introduces a unique set of challenges due to the frequent absence of necessary qualitative data. Historical texts, rich repositories of information on the interplay between literature and the environment over time, often lack the specific quantitative details essential for empirical investigations. For instance, a researcher interested in studying the representation of natural landscapes in nineteenth-century poetry selects a collection of poems from that era, encountering the challenge of the absence of quantitative data, such as specific geographical details or measurable environmental conditions. To address this limitation, the researcher might employ qualitative content analysis techniques, systematically analyzing the textual descriptions within the poems to extract qualitative insights about emotional connections between poets and nature or recurring themes related to environmental change. Overcoming this limitation requires innovative approaches for extracting relevant data from these texts, recognizing their intrinsic value in providing insights into the historical dimensions of literature and its relationship with the environment.

These limitations should not discourage but rather guide the judicious use of empirical ecocriticism. It is a valuable tool within the larger ecocritical toolbox, but it should be applied thoughtfully and in conjunction with other critical methods. While collaborating harmoniously with natural and social sciences, empirical ecocriticism remains humble, recognizing its role as one part of the larger ecocritical puzzle. Embracing diverse approaches to understanding the complex relationship between literature and the environment, empirical ecocriticism, as an ecocritical theory, remains a guiding light, illuminating new paths in the ever-evolving field of ecocriticism.

REFERENCES

Alaimo, Stacy. *Bodily Natures: Science, Environment, and the Material Self.* Indiana University Press, 2010.

Brereton, Pat, and Victoria Gómez. "Media Students, Climate Change, and YouTube Celebrities: Readings of *Dear Future Generations: Sorry* Video Clip." *ISLE: Interdisciplinary Studies in Literature and Environment*, vol. 27, no. 2, Spring 2020, pp. 385–405, https://doi.org/10.1093/isle/isaa021

Lifu, Jiang, and Scott Slovic. "Practicing Ecocriticism and the Environmental Humanities during the Age of COVID and Beyond: An Interview with Scott Slovic." *Interdisciplinary Studies of Literature*, vol. 7, no. 1, March 2023, pp. 81–98.

Małecki, Wojciech. "Experimental Ecocriticism, or How to Know If Literature Really Works?" *The Routledge Handbook of Ecocriticism and Environmental Communication*, edited by Scott Slovic, Swarnalatha Rangarajan, and Vidya Sarveswaran. Routledge, 2019, pp. 211–223.

Małecki, Wojciech, et al. "Literary Fiction Influences Attitudes Toward Animal Welfare." *PLOS ONE 11*(12), 2016, Article e0168695, https://doi.org/10.1371/journal.pone.0168695

———. "Can Fiction Make Us Kinder to Other Species? The Impact of Fiction on pro-Animal Attitudes and Behavior." *Poetics 66*, February 2018, pp. 54–63. https://doi.org/10.1016/j.poetic.2018.02.004

———. "Feeling for Textual Animals: Narrative Empathy Across Species Lines." *Poetics 74*, June 2019, 101334, https://doi.org/10.1016/j.poetic.2018.11.003

———. *Human Minds and Animal Stories: How Narratives Make Us Care About Other Species*. Routledge, 2019.

Monani, Salma, et al. "Loving Glacier National Park Online: Climate Change Communication and Virtual Place Attachment." In Handbook of Climate Change Communication: Vol. 3, edited by Walter Leal Filho et al., Springer International Publishing, 2018, pp. 63–83.

Schneider-Mayerson, Matthew. "Disaster Movies and the 'Peak Oil' Movement: Does Popular Culture Encourage Eco-Apocalyptic Beliefs in the United States?" *Journal for the Study of Religion, Nature and Culture,* vol. 7, no. 3, 2013, pp. 289–314, https://doi.org/10.1558/jsrnc.v7i3.289

———. "The Influence of Climate Fiction: An Empirical Survey of Readers." *Environmental Humanities*, vol. 10, no. 2, 2018, pp. 473–500, https://doi.org/10.1215/22011919-7156848

———. "Just as in the Book? The Influence of Literature on Readers' Awareness of Climate Injustice and Perception of Climate Migrants." *ISLE: Interdisciplinary Studies in Literature and Environment,* vol. 27, no. 2, Spring 2020, pp. 337–364, https://doi.org/10.1093/isle/isaa020

Schneider-Mayerson, Matthew, et al. "Empirical Ecocriticism: Environmental Texts and Empirical Methods." *ISLE: Interdisciplinary Studies in Literature and Environment*, vol. 27, no. 2, Spring 2020, pp. 327–336, https://doi.org/10.1093/isle/isaa022

Škobo, Milena, and Jovana Đukić. "James G. Ballard's Urban Violence Quadrilogy: An Ecocritical Approach." *DHS-Social Sciences and Humanities: Journal of the Faculty of Humanities and Social Sciences*, vol. 7, no. 3(20), 2022, pp. 87–104, https://doi.org/10.51558/2490-3647.2022.7.3.87

———. "Manifestations of 'New Age' Religions in Gated Communities of J. G. Ballard's *Cocaine Nights* and *Kingdom Come*—An Ecocritical Approach." *Palimpsest,* vol. 7, no. 14, 2022, pp. 87–100, https://doi.org/10.46763/PALIM22714100sh

———. "Urban Landscape as the Embodiment of Social and Psychological Entropy in J. G. Ballard's *Cocaine Nights* and *Super-Cannnes*." *SIC—Journal of Literature, Culture and Literary Translation*, no. 2, year 13, June 2023, https://doi.org/10.15291/sic/2.13.lc.6

Sopcak, Paul, and Nicolette Sopcak. "Qualitative Approaches to Empirical Ecocriticism . . . " In *Empirical Ecocriticism*, edited by Schneider-Mayerson et al. University of Minnesota Press, 2023, pp. 59–90.

Chapter 15

Material Ecocriticism

Reading the Nonhuman World

Gina Stamm

Material ecocriticism, in theory, is an ecocritical theory that should be easy to define. As a domain within the larger field of the environmental humanities, it emerged in publication first in 2012 in an article by Serenella Iovino and Serpil Oppermann, then in the 2014 volume edited by these same two theorists. It is a young field with a fairly stable roster of writers. On the one hand, this should make it a relatively simple object of study, but on the other, it has not had the time necessary to resolve certain questions that have arisen between different parts of the field. Iovino and Oppermann have over the years maintained a relatively stable definition and practice, as I will discuss in the first section of this essay. There are, however, two tensions that trouble the sustainability of material ecocriticism as a unique practice or discipline. One is internal: even within the 2014 edited volume, the contributors have dissonant ideas regarding the basic assumption behind the practice. This irregularity is maintained throughout the use of the term "material ecocriticism" as it has been used in publications by many authors since 2014. The other tension is between material ecocriticism and another field that preceded it and seems to overlap with certain aspects of its definition as maintained by Iovino and Oppermann: what has been called "Indigenous metaphysics." This field, under this name and related ones such as "Native American ontologies" is nearly identical to what, when developed from the Western postmodern tradition, becomes material ecocriticism. In the final part of this essay, I will argue that the work of Indigenous North and South American scholars (and a few non-indigenous scholars who have entered the field) represents a fundamental challenge to material ecocriticism. Can it continue to exist as an autonomous field of study, at least if it does not attend to the ways in which indigenous knowledge and ways of reading have preceded it and the complexities they represent?

WHAT MAKES MATERIAL ECOCRITICISM UNIQUE?

Iovino and Opperman, in the introduction to *Material Ecocriticism,* offer their own definition of the practice. They propose to "read" networks of the human and the nonhuman as a narrative or "story":

> The world's material phenomena are knots in a vast network of agencies, which can be "read" and interpreted as forming narratives, stories. Developing in bodily forms and in discursive formulations, and arising in coevolutionary landscapes of natures and signs, the stories of matter are everywhere: in the air we breathe, the food we eat, in the things and beings of this world, within and beyond the human realm. All matter, in other words, is "a storied matter." It is a material "mesh" of meanings, properties, and processes, in which human and nonhuman players are interlocked in networks that produce undeniable signifying forces. (Iovino & Opperman, Introduction)

This definition takes on board some of the fundamental suppositions of posthumanist and new materialist thought, in that material phenomena are read as agents within a network of interconnections, and the continuities between the human and the nonhuman are part of a system of signification. All matter acts on and is affected by all other matter, telling a story that is legible, if not transparent, from our position as humans. Specific to material ecocriticism within the larger spectrum of posthumanism is the vision of matter not only as agential (acting effectively on other elements within the network, what Bruno Latour would call "actants" within the collective (75), thus decentering the human, but they are also considered to be discursive entities. Material beings signify, produce meaning, and what is more, offer themselves for interpretation to the human reader.

Iovino and Oppermann take the object of material ecocriticism to be not only this discursive activity in the material world, but also in writing about materiality, at least as they describe the work being done by the scholars they have collected in the volume:

> As the approaches taken by the featured authors clearly indicate, a *material* ecocriticism examines matter both *in* texts and *as* a text, trying to shed light on the way bodily natures and discursive forces *express* their interaction, whether in representations or in their concrete reality. (Introduction)

The claim by material ecocriticism to follow the discursive nature of material in reading their textual representations may seem to pose an ethical problem. Would we not then be engaged in analysis of an appropriation of material discourse by the human authors of these texts? We could, however, see this reading of materiality in texts as a kind of second-degree material ecocriticism, taking the text being analyzed as the first degree. Writing about the material world (nature writing, fiction, etc.) would be seen not as an emanation of the author or critic's own subjectivity alone, but rather, this person acting as what Bruno Latour calls the "spokesperson" of the nonhuman (Latour 62). While not claiming transparent access to the meanings of material nonhumans' actions and discourse, they purport to attend to what the nonhuman within the collective are saying and interpreting it in a good faith effort. This is the role that Latour had attributed to the sciences (with a small "s"), who, using the prostheses offered to them by technology, can help enumerate the members of the collective of the human and nonhuman, and facilitate the "speech" of the nonhuman (67). In the world of material ecocriticism, both the poet (writer, filmmaker, etc.) and the critic find their interpretative skills useful as well. Iovino and Oppermann take the nonhuman material world to be the subject of their own writing, "reading" the narratives they see in the city of Naples (Iovino, "Storied Seas") and the ocean (Oppermann, "Ecological Postmodernism"). However, other people return to textual examples of representations of the so-called natural world with the aim of seeing how someone else has "read" a nonhuman phenomenon, for example analyzing Virgil's *Eclogues* and how Virgil's reading of nature is represented there (Rozzoni).

The other way in which material ecocriticism positions itself as useful and innovative is in the way it purports to bridge the gap between the material and the postmodern (poststructuralist) trends in criticism. As Oppermann writes:

> Theory . . . has always been a contested ecocritical issue, because it has been associated with poststructuralist interrogation of the referents of reality, which clashed with ecocriticism's central rationale of focusing on the world beyond the word. The resistance to theory in ecocriticism is a result of manifest phobias about the referents discursive nature, which is often associated with its erasure. ("Ecocriticism's Phobic Relations" 768)

This claim is somewhat hyperbolic in alleging a "phobia" within ecocriticism. While some deep ecologists may not engage explicitly with theory or hold the textuality of the "linguistic turn" in suspicion, Oppermann does not specify who she is referring to in this passage, and the "material turn" has always included theory. This has been true from its beginnings with thinkers such as Latour and Michel Serres, and both Vicki Kirby's *Telling Flesh* (1997) and Karen Barad's *Meeting the Universe Halfway* (2007)—two of the most influential texts in the new materialist and ecological discourses—engage explicitly with Deconstruction. While these last two do not explicitly claim the label of ecocriticism, they are only the most prominent of the texts also claiming to put the discursive and the material in contact. However, while these claims may be exaggerated, in his article "Comfortably Numb: Material Ecocriticism and the Postmodern Horror Film," Stephen A. Rust nuances that claim, making a distinction between practical and theoretical ecocriticism:

> This move toward material ecocriticism has served to distance us from the highly polarizing debates about postmodernism vs deep ecology and practical versus theoretical ecocriticism that characterized the field during the late 1990s and early 2000s. Rather than drawing ecocritics into separate camps—the discursive versus the real—material ecocriticism is recentering the field by bringing these two camps directly into conversation. (550)

Material ecocriticism, while not the first ecologically engaged critical theory, may serve as a bridge between those drawn to deep ecology and/or "practical ecocriticism," and more theoretically engaged practices, with its roots in theory and its attention to material and narrative. However, the extent to which this divide exists or needs to be crossed is not fully fleshed out either in Opperman or Rust's writing and remains to be explored.

ONTOLOGICAL QUESTIONS

Serpil Oppermann, in particular, has continued to develop the ideas put forward in the texts previously cited, and she specifically identifies material ecocriticism as existing in the theoretical lineage of "ecological postmodernism," which recognizes the agency or vitality of things. This indicates that she locates this divide between theory and ecocriticism as being not between textual theories of poststructuralist thought and ecological thought in general, but between ecological theories (even those generated in philosophy) and the literary field specific to ecocriticism:

> The idea that all material life experience is implicated in creative expressions contriving a creative *ontology* is a reworking of ecological postmodernism's emphasis on material process intersecting with human systems, producing epistemic configurations of life, discourses, texts, and narratives. Because ecological postmodernism perceives matter equipped with internal experience, agentic creativity, and vitality, it is important to acknowledge it as one of the roots upon which material ecocriticism constructs its theoretical premises. ("Ecological postmodernism"; emphasis mine)

In addition to the postmodern thinkers Oppermann cites, she further grounds material ecocriticism in Western theory's history with a reference to the lineage of Roland Barthes:

> What makes this interpretative horizon so interesting is not only a new reading of so-called *nonhuman objects and human subjects as co-constitutive of each other*, but also a serious challenge of logocentric thought. In this sense, material ecocriticism is mainly concerned with de-doxifying what Roland Barthes called the "doxa," or the artificially naturalized systems of meaning, by which we have developed our anthropo-logo-centric discourses. ("Theorizing Material Ecocriticism," 468; emphasis mine)

While, again, the specificity of material ecocriticism is in its practice of reading the narrative or story of matter, it occupies intellectual terrain based on an *ontological* assumption about material, living and nonliving. It requires that one attribute vitality to things. This is, on some level, an ethical commitment with implications for practitioners, and certainly that is the point of view of those who call themselves ecological postmodernists as articulated by Oppermann herself (chapter 1), for whom theoretical work is a part of a reparative ethical project. Specifically, material ecocriticism would decentre the human as a discrete entity by claiming its co-constitution with the nonhuman, and this "constitution" of the world would be a fundamental change in our understanding of its ontology.

However, in the following chapter of *Material Ecocriticism*, Hannes Bergthaller questions the use of that term: "[W]e cannot hope to simply replace [dualism], like a faulty engine, with a better *ontology*, because such semantic patterns are themselves products of social evolution" (chapter 2). Besides the metaphor of the engine being perhaps an inadequate analogy here—ontological suppositions on this level cannot merely be swapped out without requiring that the entire mechanism be restructured—such an argument reinforces the nature-culture divide by its scepticism about the "products of social evolution" by suggesting that those products of social evolution are somehow outside of the "natural world." If, as Oppermann contends, the "doxa" of anthropo-logo-centric (or bio-logo-centric) ontological discourse has been "artificially naturalized," the replacement of that doxa is an ethical commitment for her and for the ecological postmodernists she bases her worldview on, from which Bergthaller retreats. For him, "Material ecocriticism is not about the replacement of a false ontology with a true one—rather, it offers a *redescription* of the world" (chapter 2). It is understandable to question both our ability to change our assumptions about the ontology of the world and the efficacy of that change on the world "outside of the system of science." However, it introduces an incoherence into the field. What are the assumptions necessary to be a "material ecocritic"? Is it enough to write *as if* material has agency and narrative power? Is agency merely a metaphor?

This question of ontology has not been settled among the people who claim the practice of material ecocriticism. On the other extreme to Bergthaller's suggestion of a "redescription of the world" with no fundamental changes in ontological assumptions is the place of spirituality or what David Tagnani calls "mysticism," proposing:

> [A]new trajectory for material ecocriticism catered on the concept of *ecomysticism,* which denotes both the experiential and literary components of a wholly material mysticism. But why would any scholar want to move towards mysticism? In short, by synthesizing materialism and mysticism, ecomysticism facilitates the interchange between one of the most prominent trends in ecocriticism (the so-called "material turn") and one of the most ubiquitous tropes of nature writing (the quasi-religious, awe-struck reverence so typical of that genre). Further, this oxymoronic coupling may have implications outside of academic theorizing: it points to a fundamental component of the affective relationship between human and place, helping to fulfil the potential of material ecocriticism to make ethical meanings recognizable. (23)

Tagnani still backs away from claiming a spiritual dimension to his work, rejecting what he calls "disembodied spirit or pantheistic life-force" (24). What is more concerning about this proposal is, however, that it puts forth a version of "mysticism," or a rehabilitation of the affective and spiritual aspects of a relationship to nature that theory may shy away from, but his reference for that remains solely within the Western canon. And this is indicative of a larger external challenge to material ecocriticism: that of its relationship to non-Western bodies of knowledge and belief. While it shares this challenge with many other Western disciplines, it is a particularly striking case, as will be discussed in the final section of this essay.

MATERIAL ECOCRITICISM AND INDIGENOUS KNOWLEDGE

The practice of material ecocriticism and the ontological presumptions of posthumanism and "new" materialism are only new in the way they have been developed in the Western scholarly canon. In general, they have been criticized for being a belated recognition of what Indigenous beliefs and practices have long acknowledged, which now gain "legitimacy" in the academy when it comes from Western science or philosophy. Recognition of this need has come even from prominent Western thinkers, such as Rosi Braidotti in *Posthuman Feminism* (134). Ecocriticism as a larger field has welcomed the contributions of Indigenous scholars and authors. *The Ecocriticism Reader,* back in 1996, contained essays by Native American writers, and one of them, Leslie Marmon Silko, has frequently been the subject of ecocritical writing.

Well-known Indigenous science studies scholar Kim Tallbear has alluded to the ways in which new materialist and posthuman thought are recreating discourses that have long existed in Native communities:

> I read . . . that we are all of us—humans and nonhumans—a networked set of social-biological relations. [Dorion Sagan] calls us "interspecies communities." That resonates with what Vine Deloria, Jr. called an "American Indian metaphysic." And by "metaphysics" Deloria meant a "set of first principles we must possess in order to make sense of the world in which we live." I would have to include in an interspecies community or networked set of social-biological relations living beings that are both material and *immaterial,* and therein is a key difference.

Tallbear at the time, and in later work, has distinguished between Indigenous and Western multispecies discourse by critiquing the biocentrism of Western ecological thought. She extends the idea of animation to inanimate and elemental entities, citing a 1911 text: "all creation and . . . every creature. . . . The tree, the waterfall, the grizzly bear, each is an embodied Force, and as such an object of reverence" (Ibid.); she stipulates that this particular description is situated within the Lakota/Dakota tradition and does not represent a claim to universal "Indigenous" truth. Environmental humanities have begun to acknowledge the agency of the "nonliving" (in particular Jane Bennett), this approach only heightens the need for acknowledgement of and exchange with Indigenous knowledge production.

This is particularly striking in the case of material ecocriticism, whose premises of "storied matter" and the interconnections of the human and nonhuman echo the writing of a number of Indigenous thinkers. For Kristen Arola, "humans and nonhumans have relations with one another and always have . . . part of these relations includes communication and unknowability" (389), which closely approaches territory staked out for itself by material ecocriticism. One of the contributions to the volume *Decolonial Conversations in Posthumanism and New Material Rhetorics*, "Listening Otherwise: Arboreal Rhetorics and Tree-Human Relations" (Helmut Pflugfelder and Kelly) asks how we might be attentive to the articulations of plant systems, how this once-controversial idea has gained more widespread acceptance and development in the Western academy, and how different TEKs (traditional environmental knowledges) imagine those systems: "where Euro-Western science would refer to a localized ecosystem or habitat, a TEK might recognize the complex entanglement of beings infused with spirit, inhabiting a specific place, and possessing unique agencies and capacities" (3). They also invoke Gabriela Raquel Ríos's assertion that "Indigenous relationality, and thus Indigenous materiality, begins from an acknowledgment that humans and land are co-constituted" (75), which brings us back to the way in which Serpil Oppermann articulates the assumptions underlying material ecocriticism.

This is not to say that material ecocriticism is only a copy of the "traditional environmental knowledges," that it merely reproduces Indigenous knowledge, or that traditional knowledge systems have been doing material ecocriticism all along. However, while it may be a new discipline within the Western academy, material ecocriticism has to acknowledge the ways in which it touches on subjects of traditional knowledge, without extractively mining these knowledge systems for the benefit of that same academy. The necessary difficulty of negotiating the relationship between all of the posthuman/new material disciplines and Indigenous knowledge is eloquently articulated in the introduction to *Decolonial Conversations*, where Jennifer Clary-Lemon and David M. Grant explain:

> Projects involving the ontological and extradiscursive in rhetoric can only be improved by an accountability to their own complicity in the settler colonial academy, one that proceeds as a global network of knowledge workers designed to further extractive and exploitative systems. . . . Our task, then, is not one of finding analogies between the fields, nor of absorbing, adopting and transposing decolonial theory and praxis into the terms of new materialism or the posthuman. (1)

If we believe that those practicing material ecocriticism do not want to perpetuate colonialist and ethnocentric gestures in the academy, we have to take seriously the question of how to negotiate our relationships with these Indigenous ways of thinking and articulating the relationships between the human and nonhuman.

CONCLUSION

Material ecocriticism, as a young ecocritical theory, has the benefit of calling academic attention to the narrative and agential power of nonhuman nature. It also gives writers of literature credit for their own "readings" of nature by taking these portrayals seriously as the object of materialist ecocritical study. The discipline puts critics in the position of being the "spokesperson" for nonhuman agents in the kind of "politics of nature" proposed by Bruno Latour, where he assigned that role to scientists in the process of taking all living beings into account. However, if material ecocritics are to take up that mantle, they will have to respond to the dual tensions I have indicated in this chapter. First, what are their ontological assumptions? Does the nonhuman have agency we interpret, or do we write "as if" that agency exists, as Bergthaller claimed? Added to that challenge, how will material ecocriticism acknowledge and respond to indigenous knowledge systems going forward, without, however, appropriating this thought in a gesture perpetuating colonial harm? Is there room to be made for a system of knowledge based on Western philosophical theory but sharing many principles with older non-Western beliefs? What accommodation between these systems is possible in a world where the Western academy has a history of hegemonic domination that only recognizes the validity of other beliefs and knowledge—especially those with a spiritual dimension—once they have been validated by that academy? These questions of ethical engagement with and recognition of non-Western knowledge, as well as the ontological assumptions at work in the practice, are what face those who wish to engage in material ecocriticism going forward.

REFERENCES

Arola, Kristin. "My Pink Powwow Shawl, Relationality, and Posthumanism." *Rhetoric Review*, vol. 38, no. 4, 2019, pp. 375–401.

Barad, Karen. *Meeting the Universe Halfway: Quantum Physics and the Entanglement of Matter and Meaning*. Duke University Press, 2007.

Bergthaller, Hannes. "Limits of Agency: Notes on the Material Turn from a Systems-Theoretical Perspective." *Material Ecocriticism*, edited by Serenella Iovino and Serpil Oppermann, Indiana University Press, 2014, E-book ed., chapter 3.

Braidotti, Rosi. *Posthuman Feminism*. Polity, 2022.

Clary-Lemon, Jennifer, and David M. Grant. "Working with Incommensurable Things." *Decolonial Conversations in Posthuman and New Material Rhetorics*, edited by Jennifer Clary-Lemon and David M. Grant. Ohio State University Press, 2022, pp. 1–21.

Helmut Pflugfelder, Ehren, and Shannon Kelly. "Listening Otherwise: Arboreal Rhetorics and Tree-Human Relations." *Decolonial Conversations in Posthuman and New Material Rhetorics*, edited by Jennifer Clary-Lemon and David M. Grant. Ohio State University Press 2022, pp. 67–91.

Iovino, Serenella, and Serpil Oppermann. "Stories Come to Matter." *Material Ecocriticism*, Indiana University Press, 2014, E-book ed., Introduction.

———. "Theorizing Material Ecocriticism: A Diptych." *Interdisciplinary Studies in Literature and Environment*, vol. 19, no. 3, 2012, pp. 448–475, https://doi.org/10.1093/isle/iss087.

Kirby, Vicki. *Telling Flesh: The Substance of the Corporeal*. Routledge, 1997.

Latour, Bruno, and Catherine Porter. *Politics of Nature: How to Bring the Sciences into Democracy*. E-book ed., Harvard University Press, 2009. E-book.

Oppermann, Serpil. "Ecocriticism's Phobic Relations with Theory." *Interdisciplinary Studies in Literature and the Environment*, vol. 17, no. 4, 2010, pp. 768–770.

———. "From Ecological Postmodernism to Material Ecocriticism: Creative Materiality and Narrative Agency." *Material Ecocriticism*, edited by Serenella Iovino and Serpil Oppermann, Indiana University Press, 2014, E-book ed., chapter 2.

———. "Storied Seas and Living Metaphors in the Blue Humanities." *Configurations*, vol. 27, no. 4, 2019, pp. 443–461, https://doi.org/10.1353/con.2019.0030.

Rozzoni, Stefano. "From 'Pathetic Fallacy' to Affective Attunement: Reading Virgil's Eclogues through the Lens of Material Ecocriticism." *SubStance*, vol. 50, no. 3, 2021, pp. 115–132, https://doi.org/10.1353/sub.2021.0031.

Rust, Stephen A. "Comfortably Numb: Material Ecocriticism and the Postmodern Horror Film." *Interdisciplinary Studies in Literature and Environment*, vol. 21, no. 3, 2014, pp. 550–561, https://doi.org/10.1093/isle/isu083.

Silko, Leslie Marmon. "Landscape, History, and the Pueblo Imagination." *The Ecocriticism Reader: Landmarks in Literary Ecology*. Edited by Harold Fromm and Cheryl Glotfelty. The University of Georgia Press, 1996, pp. 264–275.

Tagnani, David. "Materialism, Mysticism, and Ecocriticism." *English Language Notes*, vol. 55, no. 1–2, 2017, pp. 23–31, https://doi.org/10.1215/00138282-55.1-2.23.

Tallbear, Kim. "Why Interspecies Thinking Needs Indigenous Standpoints." *Fieldsights*, November 18, 2011, https://culanth.org/fildsights/why-interspecies-thinking-needs-indigenous-standpoints.

Chapter 16

Posthuman Ecocriticism

Finding Our Way Through Collaborative Survival

Mahinur Gözde Kasurka

Human as a category is loaded with a burden that is regulated by dual logic which functions in dividing nature from culture, human from nonhuman by leading to a severe crisis in establishing the relationality of the human as a category with other living and non-living entities. To be more precise, the human as a category has been primarily established in the Enlightenment ideal by proposing a perfect model that centres *himself* in his relationship with all the others. This ideology suggests a conscious "he" figure when we look at Leonardo da Vinci's *Vitruvian Man* figure (*The Posthuman* 13). As it is clear from that figure, a certain conception of the human is proposed to make it clear who can be counted as human and who is excluded. This exclusive mindset of human centred agenda fails to embrace each human to the same degree. In other words, the model that we see in Leonardo da Vinci's endeavour requires to have certain features which is a point that posthumanist thinker Rosi Braidotti reflects by stating: "the allegedly abstract ideal of Man as a symbol of classical Humanity is very much a male of the species: it is a he. Moreover, he is white, European, handsome, and able-bodied" in *The Posthuman* (24). Thus, the categorical reference of the human sets a limit by leaving out some others outside as they are not human enough. Posthumanist agenda strives to criticize this discriminatory logic of human centred perspective as humanism is not an ideology working for the betterment of each and every human to the same degree. There is a strong division between the privileged ones and the others. Here, we need to state that there is a multi-layered process in the discriminative agenda of humanism. To be more precise, a certain conception of the human is set as the criterion within humanist vantage point. It otherizes both some nonprivileged humans and nonhumans. This close group of privileged humans also constitutes an Anthropocentric division together with human's conceptualization as an enclosed entity that separates themselves from the others. Francesca Ferrando emphasizes the same process by underlining the significance of "the human as a self-defying agent" in the conception of Anthropocentrism (*Philosophical Posthumanism* 104). In this respect, it becomes crystal clear that the humanist agenda does not only work against nonhuman, but it also excludes some humans as they fail to meet the standards set for them. Hence, the ones who fail to meet the requirements of the human as a category constitute the ones that are not privileged enough.

Posthumanism draws attention to the exclusive agenda of humanism which presupposes a fixed category of the human by belittling human's relationality with nonhuman others. In line with this, human's categorizing *himself* as a distinct entity that has cut all the bonds *he* has with nature turns out to be a problematic issue since it aims to dissociate human and nonhuman ontological spheres from each other. Accordingly, posthumanist agenda does not only criticize the idea of the human as a universal model, but it also aims to highlight the enmeshed network of agents by negating exceptional human superiority. This stance requires to be aware of the binary logic that works segregating human from all the others which resonates a failed ideology since human is a living entity that has a complex network of relations with all the other living and non-living entities. Posthumanist agenda undermines the idea of human as a separate ontological category by binding human and nonhuman, nature and culture, matter and text to each other. In so doing, in the forthcoming part of this chapter, I will lay bare human's relationality with nature on a horizontal plane by undermining the totalizing discourse of human centred agenda. With an end to portray the transformation moving from nature/culture towards *naturecultures*, this chapter intends to shed light on posthuman ecocriticism by considering agentic potentialities of the matter. Since both posthumanism and material ecocriticism morph into posthuman ecocriticism, I will briefly touch upon these theoretical frameworks in order to highlight how they *intra-act* with each other by embodying ecological awareness in Karen Barad's sense of the term. In view of this tendency, I argue that posthuman ecocriticism as a branch of ecocritical theory heralds the interconnectedness of all life on the planet by erasing ontological stability of biological categories. This enmeshed network of relations with human and nonhuman others in the age of digital capitalism has a double coded standpoint in posthumanist ecocritical vision: Firstly, it does not function in belittling the irreversible ecological destruction all around us. Secondly, in dealing with this predicament, posthuman ecocriticism aims to offer a hopeful ground as Anna L. Tsing points out with her reference to *matsutake mushroom* that struggles to enable a space for themselves to live in the ruins of capitalism (*The Mushroom at the End of the World*). Following from this, I acknowledge *matsutake mushroom* as an embodiment of posthuman ecocritical reflection as they both portray the destructive mechanism of digital capitalism together with affirmative transformation. In tune with these considerations, this chapter suggests that posthuman ecocriticism reserves some space for hopeful visions enmeshed within dystopian chaos of the presence by stressing the need to create new stories away from despair as Haraway finds the solution by making *kin* beyond reproduction (*Staying with the Trouble*).

In forming new *kinship* relations with non-human others, posthuman ecocriticism offers a new acknowledgement of subjectivity "as an expanded self whose relational category is not confined within the human species but includes non-anthropomorphic elements" in Braidotti's words ("A Theoretical Framework for the Critical Posthumanities" 42). It is through this understanding of subjectivity that posthuman ecocriticism indicates a hybridization which functions in affirmative blurring of boundaries by re-thinking of agency on a non-hierarchical frame of reference.

With an aim of explicating hybridity of posthumanist perspective, I will manifest the idea that acknowledgement of nature as a stable, untouched entity is a failed attempt in the forthcoming part. After briefly clarifying why nature/culture division, which is established as an outcome of modernity driven logic, has collapsed, I will demonstrate the ineradicable relationality between human and nonhuman from an ontological perspective.

POSTHUMANIST CRITIQUE OF NATURE/CULTURE BINARISM

Nature/culture division has been consolidated in the Enlightenment logic, and there is a general tendency of linking the establishment of duality-driven boundaries with Renaissance period. Rosi Braidotti also marks this point in history when she foregrounds her own critical engagement with human centred duality. Braidotti turns our attention to the violence exerted against the others of the discourse and its relationality with human centred agenda. She reflects:

> Environmental theory stresses the link between the humanistic emphasis on Man as the measure of all things and the domination and exploitation of nature and condemns the abuses of science and technology. Both of them involve epistemic and physical violence over the structural 'others' and are related to the European Enlightenment ideal of reason. (48)

When nature is considered as a stable entity that is waiting for humans to come and conquer, it becomes obvious that any source of exploitation becomes ordinary. Enlightenment logic contributes a lot to the idea of humans as a self-enclosed entity. Yet, it is possible to trace the beginning of this inclination of exploiting any natural sources for the sake of central human back to ancient Greek society and Christian teachings later. In ancient Greek society, there is a strict division between the privileged human and others in the categories of *bios* and *zoe*.

Bios illustrates an acknowledgement of the one who is counted as a citizen in ancient Greek society whereas *zoe* is devoid of this exclusive agenda and it also includes the ones who are not protected within the protective cloak of *bios*. By showcasing this discriminative agenda of the category's *bios* and *zoe*, Braidotti also defines *bios* as "the life of humans organized in society while *zoe* refers to the life of all living beings. *Bios* is regulated by sovereign powers and rules, whereas *zoe* is unprotected and vulnerable" (*Posthuman Knowledge* 10, emphasis in the original). In her posthumanist agenda, Braidotti marks the empowering connotations of *bios* by also illustrating how *zoe* works in a hierarchical manner. Yet, what Braidotti offers here is not crying over the dualism dominated categories, in contrast she presents how the distinctions between nature/culture, human/nonhuman, etc. have been established in ancient Greek society by signposting the existence of two separate groups. Braidotti's offer can be taken as a posthumanist dictum as she suggests "life is not exclusively human: it encompasses both *bios* and *zoe* forces" (*Posthuman Knowledge* 45, emphasis mine). If we are aware of the working logic of this dichotomy, we can unlearn "the dialectical view of consciousness based on the opposition of self and others" (Braidotti, *Posthuman Knowledge* 45).

Braidotti points out *bios* and *zoe* as the first indicator of the ontological divide between self and others and this inclination also finds itself a place in Christian theology. The divide between human/animal, human/nature is apparent in *The Old Testament* in which it is possible to see that all entities apparent in nature together with animals are created for the betterment of human's life. In other words, the human has the ability to exploit these sources on earth without any limitation by not taking into consideration their right to exist. As humans hold a central position, all the others except humans do not belong to the exclusive domain and are open to any form of exploitation. In *The Old Testament*, the creation of human is described as in the following:

> And God said, Let us make man in our image, after our likeness; and let them have dominion over the fish of the sea, and over the fowl of the air, over the cattle, and over the earth, and over every creeping thing that creepeth upon

> the earth. So God created man in his own image, in the image of God created he him; male and female created he them. And God blessed them, and God said unto them, Be fruitful and multiply, and replenish the earth, and subdue it: have dominion over the fish of the sea, and over the fowl of the air, and over every living thing that moveth upon the earth. (*The Old Testament*, *The Norton Anthology of World Masterpieces* 52)

As it is clear, human's creation is followed by their right to use all the sources on the earth, which indicates a human centred perspective. Thus, I argue that the Renaissance period consolidates the idea of a certain conception of the human as modelled in Leonardo da Vinci's *Vitruvian Man,* yet the human centred position is not born in this period. Braidotti's designating this figure is telling, but we also need to be aware of the Christian theology's forming human as the one who is allowed to have dominion over all living entities of the earth.

It should also be noted that taking human as a central entity also creates some reactions in literary productions in the ancient Greek and Roman texts. Namely, the Roman poet Ovid's *Metamorphoses* is fraught with the idea of transformation between nature/culture dichotomy by relating this to Greek mythological figures. His recounting of "Apollo and Daphne" can be taken as a case in point as Ovid highlights the transformation of Daphne, a virgin into a laurel tree which is a natural entity. Apollo, as the god of reason, ends the tale by hugging this laurel tree as he is deeply in love with this natural entity. Ovid ends the story by voicing Apollo who states: "Daphne, who cannot be my wife must be the seal, the sign of all I own, immortal leaf twined in my hair as hers, and by this sign my constant love, my honour shall be shown" (20). Hence, the metamorphosis as the apparent theme throughout the whole text comes to surface in this story by binding nature and culture together. Seen in this light, Ovid's text signifies a posthumanist endeavour in his own time by literalizing posthuman hybridity.

The idea of linking nature and culture with each other instead of struggling to construct barriers between them can also be traced in the concept of *naturecultures*. Donna Haraway categorizes *naturecultures* with an end of going beyond the dichotomous logic as she exemplifies it in the human and dog relationality. She emphasizes the fact that both humans and nature have weighty impact in dog's domestication process. Iris van der Tuin explains this category as "the impossibility of separating domains such as history and biology" and adds "any biological question has an immediate historical dimension, and any historical issue is entangled with biological processes and phenomena" ("Naturecultures" *Posthuman Glossary* 269).

Apparently, posthumanist agenda suggests a nonhierarchical and nonbinary attitude in showcasing human and nonhuman *intra-action* in Karen Barad's sense of the term. Barad offers a new interpretation of the human, highlighting a becoming process that is freed from any privileges. In this line of thinking, Barad indicates "individuals emerge though and as part of their entangled intra-relating" in her book entitled *Meeting the Universe Halfway* (ix). This *intra-active* relationality of human and nonhuman paves the way for merging assumingly separate categories about which Barad also reflects: "matter and meaning are not separate elements. They are inextricably fused together, and no event, no matter how energetic, can tear them asunder" (4). This effort in transcending several boundaries in relation to matter and text can also be seen as material ecocriticism which morphs into posthuman ecocriticism with the guidance of posthuman hybridity. In posthuman ecocritical reflection, hybridity offers a new way reading nature-culture continuum by also paying attention to the destruction that anthropos has created on earth, but not giving in to despair. Instead, we can find new ways of seeing *collaborative survival* in Anna Tsing's sense of the term from a posthuman ecocritical frame. By learning to read the *stories of matsutake mushroom*, humanity will be able to unlearn positioning themselves as the only agentic force in the world.

THE ROAD LEADING TOWARDS POSTHUMAN ECOCRITICISM

Material ecocritical stance can be acknowledged as containing overtones of interconnectedness of all life on planet by negating the ways of living that have been established under the shadow of monolithic thinking. This struggle requires to unlearn the life practices that are symptomatic of anthropocentric mindset. In doing so, this article will explicate how material ecocriticism metamorphoses into posthuman ecocriticism with an end to signify the urgent need to recalibrate human's *intra-active* relationality with nonhuman others. Within this framework, Serpil Oppermann and Serenella Iovino explain this in their text entitled *Material Ecocriticism* by expressing the following:

> In material-ecocritical terms, the human agency meets the narrative agency of matter halfway, generating material-discursive phenomena in the forms of literature and other cultural creations, including literary criticism. Also, the way we interpret the world's narratives is evidently a mode of intra-action, a phenomenon emerging from the world's creativity. (9)

Hence material ecocriticism can be comprehended as showcasing the necessity of "a reconceptualization of both the idea of text (as distinct from non-textual material formations) and the idea of the world (as 'outside of text')" (Oppermann & Iovino 10). This hybridity of matter and text echoes the category of *storied matter* as Opperman also points out: "matter is endowed with creative expressions, manifesting as *storied matter*. . . . Storied matter in the posthuman moment exhibits itself in matter's overlapping biotic and abiotic components transmitted through technoscientific practices that seek to graft the technological onto the biological" ("From Material to Posthuman Ecocriticism: Hybridity, Stories, Natures" 274).

Storied matter considers closely human and nonhuman entities forming narrative agency in their *intra-active* becoming. The idea of relating narrativity as a power solely belonging to human counterparts of this world is refuted in this mindset as Başak Ağın also states in her dissertation: "the agentic power of what we consider to be non-living or abiotic cannot be disregarded" (33). This inclusive agenda reveals how living and non-living entities function equally in a process that generates meaning. It is because of this situation that the concept of *storied matter* does not presuppose a human listener for the matter to codify their own meaning. Regarding this issue, Jeffrey J. Cohen approaches stone as embodying both matter and text within itself in his book *Stone: An Ecology of the Inhuman*. In doing so, he aims to pose a distance between himself, and the kind of stone used for the purposes of the human needs. Instead, he addresses his material as the one:

> [T]hat may be hewn but has generally not been domesticated into cornerstone or sculpture, into a display of human craft. Affixed to its earthly origin as cave wall or slope, such material was called by Roman authors vivum saxum (living rock), while vivus lapis (living stone) had been detached from its source but not incorporated into a human project. Both terms were inherited into the Middle Ages and used to convey stone's enduring vivacity. (13)

By taking stone as his primary material in his critical engagement with agency of non-living entities, Cohen demystifies the polyphonic *naturecultures*. In a similar manner, posthuman ecocriticism echoes an alternative space for the voices of the unheard by stressing the hybridity on a non-hierarchical manner.

As posthuman ecocriticism denies the idea of species supremacy together with ontological stability of the status of human, it announces that we form *kinship* relations with other human and nonhuman entities of this world, which also signifies that it is not possible to assume the human as holding an *ontologically hygienic* position. Elaine Graham posits the category of *ontological hygiene* in her book *Representations of the Post/human* (2002) by indicating the idea of dividing "human from nonhuman, nature from culture, organism from machine" (35). As posthuman ecocriticism demonstrates the impossibility of drawing boundaries between several agents of this world, we need to acknowledge that we form *kinship* connections not only with human others, but also with nonhuman others. In light of this, Donna Haraway indicates a new survival strategy by being aware of the *kinship* relations we have with earth others in *Staying with Trouble* (2016). She specifies *kinship* relations of human and nonhuman inhabitants of this world as a survival strategy in these times of crisis due to ecological degradation. If we behave in cognizant of the fact that we cannot cut the bonds we have with earth others, the boundaries will dissolve as "we require each other in unexpected collaborations and combinations . . . we become with each other or not at all" (4). Against this backdrop, posthuman ecocriticism offers us a hybrid zone in which we would be aware of the *intra-active* relationality between human/nonhuman, nature/culture, matter/text, namely human and more-than-human inhabitants of this world.

This hybridity also requires us to realize that human as a category is the one who should acknowledge their guilt in the current ecological degradation. Still, the results of this predicament do not only affect the lives of human. This posthuman ecocritical reflection signifies the interconnectedness of all life on earth but does not suggest a lamentation over the loss of pristine nature. Instead, it offers taking a stance by considering the harm Anthropos has given to ecology, but in doing so we should not posit ourselves based on the idea of "shared vulnerability" as Braidotti signifies in *The Posthuman* (79). In contrast, Braidotti indicates "a transformative or symbiotic relation that hybridizes and alters the 'nature' of each one and foregrounds the middle grounds of their interaction" (79). In fashioning a new perspective by destabilizing dualism dominated epistemologies, I would epitomize *matsutake mushroom* as an embodiment of posthuman ecocritical reflection.

MATSUTAKE MUSHROOM AS AN EMBODIMENT OF POSTHUMAN ECOCRITICAL REFLECTION

Anna Lowenhaupt Tsing in her book *Mushroom at the End of the World* (2015) goes against the Enlightenment logic which assumes to attach passivity to nature. Her criticism against species supremacist viewpoint underlines the fact that "taming and mastering has made such a mess that it is unclear whether life on earth can continue" and suggests using *matsutake mushroom* as an exemplary existence that is able to flourish in the ruins of capitalism (vii). By pointing out *matsutake mushroom*, Tsing enables us to have a third space that is an entanglement of both chaos and hope at the same time. Hence, this hybrid space gives voice to nonhuman counterparts of this world as they embody tales of resistance in the worst of times.

As Tsing articulates "matsutake mushroom's willingness to emerge in blasted landscapes allows us to explore the ruin that has become our collective home" by consolidating the status of this entity as the first one that starts emerging after atomic bomb in Hiroshima (3). By drawing from this resilience, we need to learn to live on this damaged planet by implementing *collaborative survival* as a coping mechanism. Tsing's proposal of finding a space that would work for all of us is summarized in the forthcoming part:

> Matsutake are wild mushrooms that live in human-disturbed forests. Like rats, racoons, and cockroaches, they are willing to put up with some of the environmental messes humans have made. Yet they are not pests; they are valuable gourmet treats—at least in Japan, where high prices sometimes make matsutake the most valuable mushroom on earth. Through their ability to nurture trees, matsutake help forests grow in daunting places. To follow matsutake guides us to possibilities of coexistence within environmental disturbance. This is not an excuse for further damage. Still, matsutake show one kind of collaborative survival. (4)

By learning from their survival skills, posthuman ecocritical mindset postulates making space for liveability despite ecological destruction. This view does not aim to silence the voices of the damaged planet, in contrast it offers human and nonhuman counterparts an urgent need to adapt their skills in finding a way of collaborative existence. This urgent need can be taken as the last call that we -as human dwellers of this planet—need to listen to the open space of collective resistance. In so doing, a new form of subjectivity is born that is freed from Oedipal links of human centred discourse which is indicative of affirmative blurring of the boundaries. Since posthuman ecocritical reflection does not deny the fact that humanity has created a mess on earth that it is not possible to undo the impacts of anthropocentrism anymore. Yet, it does not mean that vulnerability would lead to a silent acknowledgement of death. By learning more from the way matsutake mushroom copes with the ruins of the planet, we need to elicit new directions by highlighting the fact that:

> Staying alive—for every species—requires livable collaborations. Collaboration means working across difference. Without collaboration, we all die. . . . We change through our collaborations both within and across species. . . . The evolution of our 'selves' is already polluted by histories of encounter; we are mixed up with others before we even begin any new collaboration. (Tsing 28–29)

Under the light of posthuman paradigm, demarcating various actors from each other becomes a futile attempt. It becomes obvious that human and non-human entities, biotic and abiotic actors are in constant dialogue with each other. This non-hierarchical conversation brings with itself a posthuman ecocritical affirmative hope that becomes actual under the light of *matsutake mushroom*. As they emerge, we can find some other ways of surviving within the chaos that we ourselves have created. Otherwise, "How else can we account for the fact that anything is alive in the mess we have made?" (Tsing viii).

As a result, posthuman ecocriticism manifests hybrid zones based on the enmeshment of culturally produced and naturally born together with ecological contamination. Regarding this issue, I want to conclude by underlining that posthuman predicament does not depict a dystopian chaos of necro politics, instead it manifests a cartography by fusing organic and inorganic based on agentic capacity of the matter and meaning. As Anna Tsing states: "we don't have choices other than looking for life in this ruin" (6). Hence, finding ways of *collaborative survival* might work as a guide in illuminating both the possibilities of life and the inevitability of destruction we have generated on earth.

To conclude, posthuman ecocriticism makes it possible to imagine affirmative horizons in environmental humanities by highlighting the existence of two opposite situations at the same time: Firstly, it is not possible to ignore the ruined presence of environment. Secondly, we can still cling on to life by learning more from nature as exemplified in matsutake mushroom. Posthuman ecocriticism paves the way for hopeful horizons as an ecocritical theory that environmental humanities will make use of to embrace both hope and despair together. The posthuman dimension of ecocriticism that is discussed all throughout this chapter as a ground for ecocritical theory will not give in to the end of the world scenarios. In contrast, posthuman ecocriticism will enable humanity to create new narratives, new stories to undo discriminative and hierarchical mindset within the field of environmental humanities by showing examples of *collaborative survival* across multiple species.

REFERENCES

Ağın, Başak. "Posthuman Ecologies in Twenty-First Century Short Animations." 2015. Hacettepe University. PhD Dissertation.

Barad, Karen. *Meeting the Universe Halfway: Quantum Physics and the Entanglement of Matter and Meaning*. Durham: Duke University Press, 2007.

Bible. *Norton Anthology of World Masterpieces.* Ed. by Maynard Back, 1965.

Braidotti, Rosi. *The Posthuman*. NJ: Polity Press, 2013.

———. *Posthuman Knowledge*. Polity Press, 2019.

———. "A Theoretical Framework for the Critical Posthumanities." Vol. 36, Issue 6. *Theory, Culture and* Society, 2019.

Cohen, Jeffrey J. *Stone: An Ecology of the Inhuman.* University of Minnesota Press, 2015.

Graham, Elaine L. *Representations of the Post/Human: Monsters, Aliens and Others in Popular Culture*. Manchester University Press, 2002.

Haraway, Donna. *Staying with the Trouble: Making Kin in the Chthulucene*. Duke University Press, 2016.

Iovino, Serenella and Serpil Oppermann. *Material Ecocriticism*. Indiana University Press, 2014.

Oppermann, Serpil. "From Material to Posthuman Ecocriticism." *Handbook of Ecocriticism and Cultural Ecology*. Ed. by Hubert Zapf. 2016, pp. 273–294.

Ovid. *Metamorphoses*. Trans. by Horace Gregory. The Viking Press, 1958.

Tsing, Anna Lowenhaupt. *The Mushroom at the End of the World*. Princeton University Press, 2015.

Tuin, Iris van der. "Naturecultures." *Posthuman Glossary*. Eds. Rosi Braidotti and Maria Hlavajova. Bloomsbury Academic, 2018, pp. 269–270.

PART III

Traversing Disciplinary Boundaries

Chapter 17

Blue Humanities

Oceans as History, Matter, and Imagination

Aina Vidal-Pérez

The sea is neither a void nor an abyss. Nor is it a static, obvious object; rather, it is ambiguous, transformative, and hybrid. Viewed from the distance of a shore, ship, or plane, the sea challenges facile representation. Charged with emotion and intensity, the sea has historically been a source of poetic recreation, a paradigm of the sublime, an antagonistic force. It is symbol and matter, natural and social. Simultaneously, the ocean has played a central role in the movements of modern societies. It was and continues to be a fundamental space for the accumulation of capital, yet it is, more than ever, a limitation on capitalist reproduction. Getting to know the ocean requires facing its unmanageability and exploring paths of imagination. Though it cannot be grasped, or precisely because of this, it is "one of the most 'universal' symbols in literature," as Jonathan Raban puts it (3). Notwithstanding, the study of oceans has remained on the outskirts of literary studies. Indeed, a focus on land dominated academia until the turn of the millennium. Progressively bridging this forgetting of the seas—what Margaret Cohen calls "hydrophasia" ("Literary" 658)—the sea has enjoyed increasing attention within the humanities in general, and in literary studies in particular, becoming a convincing macro-object of analysis.

This essay critically reviews the emergence of the so-called Blue Humanities as a suitable eco-concept to understand the relevance and topicality of an environmental approach to the oceans. Aiming to problematize the widespread perceptions of the oceans and the tensions and circumstances such perceptions sustain, this chapter argues for an ecocritical conception of the oceans as domains in which history, matter, and imagination problematically confront.

VEERING FROM LAND TO SEA

While the origins of this oceanic trend are complex, scholars agree that the so-called maritime turn shares certain contextual triggers. The spatial turn in post-1970s led to understanding the social, historical, material, and imaginative dimensions of space, broadening discourses of globalization and loosening traditional, nation-based approaches. The rise of postcolonial and diaspora studies largely contributed to shedding light on how the histories of colonial resistance took place in transoceanic contexts in the face of modern imperialism. New perceptions of ethnicity and race settled the oceans as passages of fluidity and creolization. Globalization is also key to the maritime turn: in a society of diverse, long-distance connections, the vast majority of operations are still conducted by sea. Specifically, within the literary field, the global turn of the last two decades appears to provide an ideal theoretical background for the shift to vast, transnational spaces, such as the oceans. Finally, in light of climate change and ensuing sea-level rises, the ecological and the environmental—hand in hand with their academic fields, ecocriticism and environmental humanities—are crucial not only for our understanding of seas, but of the planetary (Spivak 72).

Mainly developed in the scholarly circles of the Anglophone West, the discipline of Atlantic history is established as the starting point for new theorizations on oceans in response to the claims for what has been called a "new thalassology" (from the Greek *thalassos*, or "sea"). Aiming to displace the terrestrial angle, oceanic studies stand out for bringing oceans to the forefront, considering them as organizing structures and not just as mere backdrops (Mentz, *At the Bottom* 101; Ganguly, "Oceanic" 430). Paul Gilroy's *The Black Atlantic* (1993) is regarded as seminal to the shift towards the maritime. Focusing on the transoceanic movement of enslaved people, Gilroy addresses the Atlantic as a middle passage and suggests that trans-Atlantic slavery fostered a counterculture of modernity. Among other pioneers who drew attention to how the ocean's imaginary and material dimensions should be studied as organizers of histories, relations, identities, politics, knowledge, and cultures, but also literatures and literary forms and traditions, we may include Édouard Glissant's *Poetics of Relation* (1997), on Caribbean culture and identity creolization. Worth mentioning is the 2000 volume *The Many-Headed Hydra: Sailors, Slaves, Commoners, and the Hidden History of the Revolutionary Atlantic*, by Peter Linebaugh and Marcus Rediker, devoted to the history of proletarianization, exploitation, and resistance in the Atlantic in the making of the modern world. Likewise, *The Social Construction of the Ocean* (2001) by Philip E. Steinberg links the construction of oceanic spaces with modern ideologies of capital. Margaret Cohen's *The Novel and the Sea* (2010) unsettles the idea that the history of the novel is bound to nations and territories and rethinks the genre from a maritime perspective. In a similar vein, other scholars, including Steven Mentz on the Atlantic ("Toward"; *At the Bottom*) and Elizabeth DeLoughrey on the Caribbean and the Pacific (*Routes*), have expanded the question of how the sea affects literary frameworks, forms, genres, and traditions. Furthermore, Isabel Hofmeyr ("The Black"; "The Complicating") has argued for the distinctiveness of the Indian Ocean based on its cultural and political relations.

All these contributions have been shaping a new field of study known as "Blue Humanities" or "Blue Cultural Studies" (Mentz, "Toward" 997), "Oceanic Studies" (Blum 670), "Critical Ocean Studies" (DeLoughrey, "Submarine" 37), or "Humanist Oceanic Studies" (Price 45). The manifold ways of approaching oceans have given rise to a wide range of terminology and even to subfields, depending on the focus of interest, such as "Terraqueous Ecocriticism" (Brayton 99), "Terraqueous Transregionalisms" (Ganguly, "Oceanic" 432), "Aquatic Environmentalism" (Alaimo, "Jellyfish" 140), and "Blue Ecocriticism" (Dobrin 8). The preeminence of the term "blue" seeks to displace the "green" in environmental discourse, however limited both metaphors may be. The guiding principle of the field is to decentre land-based perspectives, which are very focused on "place-based identities, and the development of stable social institutions, most notably those associated with state power" (Steinberg, "Of Other" 157). Calls for alternative regionalizations dominate this new trend in literary studies, which "must consider the physical environment as a substantial partner in the creation of cultural meaning" and "literary habits of thought" (Mentz, "Toward" 1008). In other words, this new area of studies seeks to lend prominence to the sea's material element and bring into play associated concepts such as connectivity and exchange, networks and assemblages, submersion and depths, rights and sovereignty, hybridization, ecocide, and global commons, all of which must be critically revised in order to explore new epistemologies of the ocean and alternative narratives and paradigms within world history and culture.

The aim of the following sections is to trace the emergence of environmental thought within ocean studies and to propose a discussion of the implications of considering the ocean both as a conceptual object and as a historically informed material domain in literary history and environmental ideology.

OCEANIC PERCEPTIONS

Based on the core precepts of the spatial turn, voices from critical geography have addressed oceans as multilayered social constructions, emphasizing their material and historical uses and consequent representations. In his 2001 cited work, Steinberg argues that the geophysical circumstances of seas have elicited different and contradictory perceptions of the ocean, be it as a void, a space to traverse, or a realm of capitalist control and power. In this text's wake, in 2010, literary scholar Hester Blum wrote that oceanic studies had thus far failed to integrate the ocean as a real, socially experienced domain, and succinctly stated that "the sea is not a metaphor" (670). Aiming to move past the idea of the sea as an abstract entity that is described through figurative language, Blum advocates for "a practice of oceanic studies that is attentive to the material conditions and praxis of the maritime world" (670), including trade, labour, sovereignty, affiliation, rights, and citizenship. Responding to Blum's provocative alternative, Steinberg suggested in 2013 that we not "deny the importance of either the human history of the ocean or the suggestive power of the maritime metaphor" ("Of Other" 157). Steinberg proposes that to fully appreciate oceans they must be brought into the image "as wet, mobile, dynamic, deep, dark spaces that are characterized by complex movements and interdependencies of water molecules, minerals, and non-human biota as well as humans and their ships" (159). The tools to do that, Steinberg argues, are found in oceanography. However, how is literature transferring that experience? What analytical tools are available for literary studies?

Steinberg's proposal to develop an epistemology of oceans as constantly renewed, reanimated, and reconstructed by a wide variety of elements beyond the human invites us to explore oceans from an environmental perspective. As Marta Puxan-Oliva states, "the accelerated environmental damage to the oceans is a cornerstone in the definition of the Anthropocene" ("Global" 46). Let us simply observe the generalized extreme weather and rising sea levels; the great Pacific garbage patch; nuclear waste across the oceans and seas; oil spills in the Mexican Gulf; overfishing in the Sea of Japan and Patagonia; transnational strife over minerals and microbiota in the Pacific Ocean; the race for control over the Arctic as ice caps melt; the coral-reef crisis due to oceanic warming, acidification, sewage, and agricultural runoff; and the massive, extractivist tourism industry in the Caribbean and Mediterranean seas. Oceans play a key role in the climate crisis, and the extension and range of their effects are challenging in terms of perception, analysis, and representation. A key idea in this regard is Pieter Vermeulen's reminder that "the reorganization of human life in the Anthropocene also afflicts the novel" (196).

How are literary forms emerging, transforming, and adapting in their attempt to transfer the vastness both of oceans and of the climate emergency? Which aspects of genres and which poetic strategies are being deployed? How are they affiliated with literary tradition? At the crossroads of the historical and material uses and abuses of the sea and its representational and perceptual devices, we can distinguish several critical and theoretical developments of the oceanic in literature.

MATERIALISMS AT SEA

Departing from the idea of oceans as vast, powerful, and dispersed, Stacy Alaimo advocates for tracing maritime substances that reveal linkages between terrestrial humans and sea creatures. In so doing, she seeks to dispel the idea of maritime as alien to human activity. Indeed, new materialist theories emphasize how "the impossibility of an ontological divide between nature and culture are invaluable for contemplating the seawater itself" (Alaimo, "States" 489). Oceanic space is being claimed as posthuman, a hyperobject that "challenges any aspirations of unconditional anthropocentrism" (Frank 41). Alaimo has been very prolific in this approach and proposes "transcorporeality" as a "new materialist and posthumanist sense of the human as substantially and perpetually interconnected with the flows of substances and the agencies of environment" ("States" 476). DeLoughrey suggests, for Caribbean aesthetics, the notion of "sea ontologies" ("Submarine" 33) to refer to a conception of the sea as a multispecies domain in which oceanic matter is embedded in human history. In a similar vein, Serpil Oppermann, Elena Past, and Serenella Iovino are exploring the Mediterranean as a natural-cultural compound. Drawing on Alaimo's contribution, Iovino calls for an "amphibian" approach to the Mediterranean and Past states that "Mediterranean ecocriticism, with its liquid focus, responds to an ethical imperative to reconnect the flows of matter and narrative and meaning" (381). Oppermann takes up the question of seascapes as materially constituted and discursively conceptualized and asks, "What if there are narrative pathways for marine organisms to express themselves in contingent patterns of creativity?" (452). Against some postmodernist and poststructuralist dematerializing trends, clearly this current seeks to radically decentre human exceptionalism. Despite its sometimes-excessive displacement of social agency, this perspective is no doubt a fruitful tool to analyse the narrative negotiation of the oceanic in light of the environmental crisis: what narrative forms do water transformations produce? Which tropes do tides, floodings, and droughts trigger? How is the temporality of global warming transferred into the narrative? Are novelists experimenting with realist, allegorical, mythical, or speculative narratives?

Undoubtedly, this perspective could foster narratological reflection to answer these kinds of questions.

The fact is that, in the maritime domain, human exceptionalism is tightly linked to the Romantic literary and philosophical ideology by which "the natural" and "the wild" are separated from the human. Yet it is also linked to capitalist political ideology, which, by means of this separation, reinforces the nonhuman as a disposable resource at the service of human necessity. In the introduction to *Blue Ecocriticism and the Oceanic Imperative*, Sidney I. Dobrin (2021) uses the concept of "frontier mentality" as "an anthropomorphic ideology that . . . maintains a superiority and separation of human from nonhuman, avails the latter as resource, and places the 'natural' world in servitude to the human, particularly as economic stores" (16). In 2020, Alexandra Campbell and Michael Paye coordinated a special issue titled "World Literature and the Blue Humanities" for the academic journal *Humanities*. Following Jason W. Moore's understanding of capitalism as an environment-making system and his argument on the dialectical relationship between nature and society, the contributors reflect on the material and symbolic paradigms that violently organize "nature's disposable" water for capitalist and colonial purposes. Oceans are explored in terms of hydrocolonialism, hydropower, statehood, forced Western modernity, tourism, and militarism, fundamentally acknowledging stories of anticolonial, anti-accumulative local resistance.

This natural-social argument can be pushed forward by thinking about narrative theory and how this theoretical field has perceived spaces merely as a fictional domain that is therefore severed from historical and environmental context. In her 2018 study of Juan Benet's *Sub rosa* (1973) and Herman Melville's *Benito Cereno* (1856), Puxan-Oliva claims that the available tools of narratology for spatial analysis fall short when dealing with "the environmental imagination of the oceans" (3). She suggests the notion of "environment," comprehensively understood both as a multilayered, historically active force and as a stimulus for specific narrative strategies. In a similar vein, focusing on the use of narrative modes addressing the global environmental crisis, Ganguly analyses Amitav Ghosh's *The Hungry Tide* (2004). The novel is an example of what Ganguly calls "planetary realism," a narrative mode that imagines human induced environmental impacts in the present that will last beyond human survival on the planet.

OCEAN'S JURISDICTION

Overcoming the perception of sea waters as a divide between humans and nature is also key to understanding how these out-of-sight spaces are being used to conduct out-of-jurisdiction activities. In other words: it is crucial that we disrupt perceptions around the oceanic offshore and that we understand the appropriation of global commons. Firstly, the offshore is related to a perception of the ocean as "outlaw," a space to discard and forget the consequences of capitalist accumulation. Through the image of a geography hidden from the general view, the ocean is seen as the perfect space to eschew official jurisdictions and leave no trace of opaque transactions. In their work on capitalism and the sea, Liam Campling and Alejandro Colás dedicate a chapter to the offshore and define it as "an actual place and a legal fiction" (268), characterized by exception and exemption. Interestingly, Campling and Colás formulate their argument using literary lexicon and link the offshore to the utopia. Harry Pitt Schott (2020) also explores the representation of the oceanic offshore through Carlos Fuentes's *The Hydra Head* (1978) and Ian Rankin's *Black & Blue* (1997) and concludes that noir mysteries are *the* genre of the offshore: noir perfectly addresses faraway, enigmatic territories while investing in questions of power, legality, and social tension.

Secondly, the global commons designate spaces that are not administrated by specific state treaties and whose goods can be extracted lawlessly: oceans, but also poles, deserts, or airspace. By analysing Frank Schätzing's *The Swarm*, in a 2022 essay Puxan-Oliva shows how literature can facilitate and upset the dimensions and perceptions of oceans as "global environments": "wide, open spaces that are difficult to regulate, inhabit and settle due to their physical nature" (42), spaces that are internationally contested, ruled under the principles of free access, and available for manifold modes of extractivism. Similarly, other scholars perceive oceans as "a final battleground for what remains of a global commons" (Price, 46) or as "a final frontier that will no more belong to all human kind" (Ganguly, "Oceanic" 432). However, my belief is that this rhetoric of loss does not favour the possibility of politically imagining these domains whose commonality is about to disappear. In this sense, Matthew Hart examines the problematic "international" status of the oceans and considers that "what we think of as 'global' space is the product of agreements, conflicts, and compromises between national states" (4). Ocean studies, therefore, could help to articulate the ambiguous agencies of global capitalism accumulation dynamics and attend to the intermediary bodies between the global and the local, be they corporations, states, and other structures.

CONCLUSION

Blue Humanities is a prolific field of inquiry for literary studies when addressed from a more humanist approach but also from an ecological approach, as exposed in this essay. The area is particularly rich in comparative literature, which has become quite obvious after the global turn (see Roig-Sanz and Rotger) and the reemergence of world literature theories (see Helgesson, Frank, and Frydman). The study of maritime literature provides new scales, lexica, tropes, plots, and figures as well as other poetic, formal, and generic features. Coasts, archipelagos, islands, frozen seas, desert seas, and deltas are also included in the picture. As seen, a study of the poetics of oceans and how they portray the complexity of the environmental crisis could help explore how these perceptions may be transferred to the social arena. However, no matter how celebratory this new field of inquiry may be, we must remain aware of its pitfalls within the humanities to ensure that it will emerge as more than just an academic trend. This essay has shed light on the frictions between the material and the imaginary, the historical and the symbolic, in order to observe the contradictions between literary, environmentalist, and capitalist discourses. Oceans have been examined as environments in which history, matter, and imagination collide, whose complexity cannot be synthesized in a single essay or within one totalizing perspective, frame, or method. As shown, Blue Humanities has proved to be a pivotal eco-concept insofar as an ecocritical perspective of oceans can decentralize general assumptions on environments. By refraining from simply reproducing the discourse of the inevitability of globalization, this field has the potential to open compelling paths in the intrinsic nature-human relationship, to showcase oceans' ambiguity, to examine the tensions underlying their apparent vastness and wilderness, and to trouble the circumstances and understandings of such tensions.

FUNDING

This chapter was possible thanks to the R&D project 'The Novel as Global Form. Poetic Challenges and Cross-border Literary Circulation' (PID2020-118610GA-I00 / AEI / 10.13039/501100011033).

REFERENCES

Alaimo, Stacy. "Jellyfish Science, Jellyfish Aesthetics: Posthuman Reconfigurations of the Sensible." *Thinking with Water*, edited by Cecilia Chen, Janine MacLeod, and Astrida Neimanis. McGill-Queen's University Press, 2013, pp. 139–164.

———. "States of Suspension: Trans-corporeality at Sea." *Interdisciplinary Studies in Literature and Environment*, vol. 19, no. 3, 2012, pp. 476–493. https://doi.org/10.1093/isle/iss068

Blum, Hester. "The Prospect of Oceanic Studies." *PMLA*, vol. 125, no. 3, 2010, pp. 670–677. https://doi.org/10.1632/pmla.2010.125.3.670

Brayton, Dan. *Shakespeare's Ocean: An Ecocritical Exploration*. University of Virginia Press, 2012.

Campbell, Alexandra, and Michael Paye. "Water Enclosure and World-Literature: New Perspectives on Hydro-Power and World-Ecology." *Humanities*, vol. 9, no. 3, 2020, pp. 1–15. https://doi.org/10.3390/h9030106

Campling, Liam, and Alejandro Colás. *Capitalism and the Sea. The Maritime Factor in the Making of the Modern World*. Verso, 2021.

Cohen, Margaret. "Literary Studies on the Terraqueous Globe." *PMLA*, vol. 125, no. 3, 2010, pp. 657–662. https://doi.org/10.1632/pmla.2010.125.3.657

———. *The Novel and the Sea*. Princeton University Press, 2010.

DeLoughrey, Elizabeth. "Submarine Futures of the Anthropocene." *Comparative Literature*, vol. 69, no. 1, 2017, pp. 32–44. https://doi.org/10.1215/00104124-3794589

———. *Routes and Roots: Navigating Caribbean and Pacific Island Literatures*. University of Hawai'i Press, 2007.

Dobrin, Sidney I. *Blue Ecocriticism and the Oceanic Imperative*. Routledge, 2021.

Frank, Søren. *A Poetic History of the Oceans: Literature and Maritime Modernity*. Brill, 2022.

Frydman, Jason. "Oceans, Archipelagos, and World Literature." *The Routledge Companion to World Literature* (2nd ed.), edited by Theo D'haen, David Damrosch and Djelal Kadir. Routledge, 2022, pp. 442–451.

Ganguly, Debjani. "Oceanic Comparativism and World Literature." *The Cambridge History of World Literature*, edited by Debjani Ganguly. Cambridge University Press, 2021, pp. 429–457.

———. "Catastrophic Form and Planetary Realism." *New Literary History*, vol. 51, no. 2, 2020, pp. 419–453. https://doi.org/10.1353/nlh.2020.0025

Gilroy, Paul. *The Black Atlantic: Modernity and Double Consciousness*. Harvard University Press, 1993.

Glissant, Édouard. *Poetics of Relation*. University of Michigan Press, 1997.

Hart, Matthew. *Extraterritorial: A Political Geography of Contemporary Fiction*. Columbia University Press, 2020.

Helgesson, Stefan. "Geographies: Reading the Ocean." *Literature and the World*, edited by Stefan Helgesson and Mads Rosendahl Thomsen. Routledge, 2020, pp. 81–108.

Hofmeyr, Isabel. "The Complicating Sea: The Indian Ocean as Method." *Comparative Studies of South Asia, Africa and the Middle East*, vol. 32, no. 3, 2012, pp. 584–590. https://doi.org/10.1215/1089201X-1891579

———. "The Black Atlantic Meets the Indian Ocean: Forging New Paradigms of Transnationalism for the Global South—Literary and Cultural Perspectives." *Social Dynamics*, vol. 33, no. 2, 2007, pp. 3–32. https://doi.org/10.1080/02533950708628759

Horden, Peregrine, and Nicholas Purcell. *The Corrupting Sea: A Study of Mediterranean History*. Wiley-Blackwell, 2000.

Iovino, Serenella. "Introduction: Mediterranean Ecocriticism, or, A Blueprint for Cultural Amphibians." *Ecozon@*, vol. 4, no. 2, 2013, pp. 1–14. https://doi.org/10.37536/ECOZONA.2013.4.2.525

Linebaugh, Peter, and Marcus Rediker. *The Many-Headed Hydra: Sailors, Slaves, Commoners, and the Hidden History of the Revolutionary Atlantic.* Beacon Press, 2000.

Mentz, Steven. "Toward a Blue Cultural Studies: The Sea, Maritime Culture, and Early Modern English Literature." *Literature Compass*, vol. 6, no. 5, 2009, pp. 997–1013. https://doi.org/10.1111/j.1741-4113.2009.00655.x

———. *At the Bottom of Shakespeare Ocean.* Continuum, 2009.

Oppermann, Serpil. "Storied Seas and Living Metaphors in the Blue Humanities." *Configurations*, vol. 27, no. 4, 2019, pp. 443–461. https://doi.org/10.1353/con.2019.0030

Oruc, Firat. "Thalassological Worldmaking and Literary Circularities in the Indian Ocean." *Comparative Literature*, vol. 74, no. 2, pp. 147–155. https://doi.org/10.1215/00104124-9594787

Past, Elena. "Mediterranean Ecocriticism: The Sea in the Middle." *Handbook of Ecocriticism and Cultural Ecology*, edited by Hubert Zapf. De Gruyter, 2016, pp. 368–384.

Pitt Scott, Harry. "Offshore Mysteries, Narrative Infrastructure: Oil, Noir, and the World-Ocean." *Humanities*, vol. 9, no. 3, 2020, pp. 71–. https://doi.org/10.3390/h9030071

Price, Rachel. "Afterword: The Last Universal Commons." *Comparative Literature*, vol. 69, no. 1, 2017, pp. 45–53. https://doi.org/10.1215/00104124-3794599

Puxan-Oliva, Marta. "Global Narrative Environments, or the Global Discourse of Space in the Contemporary Novel." *Global Literary Studies: Key Concepts,* edited by Diana Roig-Sanz and Neus Rotger, De Gruyter, 2022, pp. 39–60.

———. "Colonial Oceanic Environments, Law and Narrative in Herman Melville's *Benito Cereno* and Juan Benet's *Sub rosa*." *English Studies*, vol. 99, no. 4, 2018, pp. 426–441. https://doi.org/10.1080/0013838X.2018.1480136

Raban, Jonathan (ed.). *The Oxford Book of the Sea*. Oxford University Press, 2001 [1992].

Roig-Sanz, Diana, and Neus Rotger, editors. *Global Literary Studies: Key Concepts*. De Gruyter, 2022.

Spivak, Gayatri. *Death of a Discipline*. Columbia University Press, 2003.

Steinberg, Philip E. "Of Other Seas: Metaphors and Materialities in Maritime Regions." *Atlantic Studies*, 2013, vol. 10, no. 2, pp. 156–169. https://doi.org/10.1080/14788810.2013.785192

———. *The Social Construction of the Ocean*. Cambridge University Press, 2001.

Vermeulen, Pieter. "'The Sea, Not the Ocean': Anthropocene Fiction and the Memory of (Non)human Life." *Genre*, vol. 50, no. 2, pp. 181–200. https://doi.org/10.1215/00166928-3890028

Chapter 18

Nuclear Humanities

Envisioning the Nuclear as a Societal and Ecological Value

Inna Sukhenko

Nowadays the study of nuclear energy related issues and its impact on the environment is getting a more and more an important topic among humanities scholars, incorporating the various forms of media, storytelling, and creative expression, from literature and film to music and movement to data visualization, to clarify the place of nuclear energy in exploring human interaction and interconnectedness with the environment. In this concern, being related to the study of the environmental studies via ideas and practices for preventing the depletion of natural or physical resources, *nuclear humanities* can be regarded among the range of eco-concepts, clarifying the vital interactions between organisms and their environment.

The debate about nuclear energy is complex and multifaceted, and there are valid arguments on both sides. It is important to consider all aspects of the issue, including safety, cost, environmental impact, and energy security, when making decisions about the use of nuclear energy. Despite that, studying the humanities' perspectives on 'the nuclear' for conceptualizing nuclear energy as a societal value within communicating the environment, sustainability, biodiversity, and planetary well-being, enhances us to believe in the need and significance of *nuclear humanities* among eco-concepts, contributing to clarifying the actors' interaction and co-existence within the environment and communicating the strategies of providing affordable energy to all who lack access, according to Sustainable Development Goals (Nuclear Energy and Sustainable Development).

At that right moment when humanity faced the discovery of radiation (Gager 762; Campos 3) in the 1890s, considered to be "a paradigm shift" (Kuhn 150), as well as further nuclear-related issues, the nuclear appeared to be within the range of emotions, potentially interpreted in various ways based on the contents and contexts. On the one hand, emoting the nuclear means experiencing intense emotions related to the potential dangers and devastating impacts of nuclear technology and nuclear weapons, which mainly refer to framing such feelings of fear, anxiety, and grief over the potential harm caused by nuclear accidents, nuclear disasters, or nuclear warfare. But on the other hand, emoting nuclear power demonstrates the range of emotions, associated with the benefits of nuclear technology, and can be represented by such feelings as excitement, hope, and progress. Such variety of the emotions inevitably meets the necessity to study the specificity of individual as well as societal attitudes, experiences, and behaviours, related to the nuclear, and making the nuclear a debatable societal value, from the perspective of humanities within debating the long-term impacts of the nuclear on human health and the environment, well-being and future energetic scenarios.

Humanities not only contribute to understanding the complexities of the nuclear age, including the role of nuclear technology in shaping contemporary societies and cultures, the environmental and social impacts of nuclear energy, and the ethical and moral considerations of nuclear weapons and their proliferation, but also play a critical role in studying the various tools of translating the knowledge on radioactivity and nuclearity to the public and vice versa—the response of the public to the agenda of humanities on the nuclear, while being based on the contexts and situations.

FRAMING *NUCLEAR HUMANITIES*

The term *nuclear humanities* is a relatively recent one, and there is no single mention of it that can be deemed its first use. Since the discovery of radioactivity the humanities' perspectives on the nuclear cover the various aspects of examining the impact of nuclear technologies on human society in all its aspects, including literature, art, history, philosophy, sociology, politics, and anthropology, etc., which makes *nuclear humanities* an interdisciplinary field of study that explores the cultural, social, ethical, and political dimensions of nuclear technology and its effects on humanity.

The term *nuclear humanities* as a concept and an academic field emerged in the twenty-first century as a result of the increasing needs for interdisciplinary approaches to understanding nuclear technology and its role in shaping the human energy-driven society. *Nuclear humanities* explore the ethical and moral implications of nuclear technology and seek to provide a meaningful dialogue and informed decision-making on nuclear energy and nuclear weapons, which encourages the interdisciplinary research initiatives, engagement with the public, and advocacy for nuclear disarmament and energy transition to sustainable alternatives (Verma iii). It is *nuclear humanities* that is the field, seeking to understand the complex relationships between humans, nuclear technology, and the environment.

The emergence of *nuclear humanities* can be traced back to the beginning of the 1980s when interdisciplinary approaches to studying nuclear technology and its impact on the human society started gaining momentum. It is the late nineteenth century, defining a growing interest in radioactivity and its implications for understanding the natural world, when scientists such as Marie Curie and Ernest Rutherford conducted ground-breaking research that paved the way for nuclear physics, while others explored the potential uses of radioactivity in medicine, industry, and warfare (Malley 213). The discovery of radioactivity in the late nineteenth century faced the public responses public via debating nuclear energy related issues in different contexts, stemming from growing awareness of the dangers of radiation, particularly among workers in the nascent nuclear industry. Industrial accidents and radiation poisoning raised concerns about the long-term health effects of nuclear energy, and some scientists and social reformers began to question the wisdom of pursuing nuclear power.

While the fields of literature, history, philosophy, and other humanities did not yet explicitly engage with these issues in relation to nuclear technology, they would grow to play an important role in shaping public debate and policy on nuclear issues in the debates that took place then, but *nuclear humanities* as a field of the interdisciplinary study of nuclear issues can be traced to that early twentieth century interest in radioactivity from a humanistic perspective. One of the brightest examples of a humanistic perspective on the nuclear is the literary writing by H.G. Wells (1866–1946), who wrote about the potential of atomic power before it had even been discovered. His novel *The World Set Free* (1914) depicts a world under transformation by a powerful new energy source, later revealed to be atomic power, which leads to war and destruction. Wells used his writing to explore the moral and ethical implications of this new energy source, and his work influenced subsequent literature and debates on nuclear issues.

Being engaged in early discussions on nuclear technology, it is the mathematician and philosopher Bertrand Russell (1872–1970), who wrote in his book *Principles of Mathematics* (1915) about the potential danger of atomic energy while emphasizing that when vast possibilities of devastation will be opened, the internal energy of atoms is unlocked, by stating that:

> We do not quite know what may be the effects of letting loose great floods of radio-activity . . . I do not say that this will happen if atomic energy is employed in war; no one knows yet whether it will happen or not. But there is a risk that it may happen, and if it does repentance will come too late. (Russell 683)

Psychologist Carl G. Jung also addressed the psychological impact of the potential use of atomic weapons in his 1945 essay *After the Catastrophe,* in which he reflects on the existential threat posed by nuclear weapons and the need for humanity to confront its darker impulses (Feldman 107). While *nuclear humanities* did not yet exist as a formal field of study before 1945—when the first atomic bomb test, code-named Trinity, was conducted on July 16, 1945, in Alamogordo, New Mexico, United States, by a team of scientists and engineers, including Robert Oppenheimer, Enrico Fermi, and US government officials, as part of the Manhattan Project (1942–1946) (Rhodes 15)—these early works and discussions laid the groundwork for the interdisciplinary exploration of nuclear issues in the years that followed, considered as *the Nuclear Anthropocene* (Carpenter 117).

This successful test and the further detonation of the atomic bombs in Hiroshima and Nagasaki in August 1945 marked not only a major milestone in the development of the atomic bomb, but also a turning point in the way how the society considered the nuclear issues. Before the dropping of atomic bombs on Hiroshima and Nagasaki in 1945, the focus of nuclear discussions was primarily scientific and technical but with a few instances in which the humanities intersected with nuclear issues. It was in the aftermath of World War II, when *nuclear humanities* emerged as a field of study, allowing social scientists, philosophers, writers, and artists to explore the complex and far-reaching implications of nuclear technology and its controversial use—as a source of energy and as a source of atomic weapon. One of the most notable contributions to *nuclear humanities* then was made by the philosopher and political theorist Hannah Arendt, who wrote extensively about the moral and political implications of atomic weapons. Arendt's book *The Origins of Totalitarianism* (1951) argued that the destruction, caused by World War II, and the use of atomic weapons raised profound questions about the nature of power, humanity, and responsibility (Arendt 490).

One of the prominent scholars and philosophers, who significantly contributed to the field of *nuclear humanities* and launched nuclear criticism as a critical literary field, was Jacques Derrida (1930–2004), whose work focused on the language and discourse surrounding nuclear weapons, particularly in the political realm (Derrida 20–21). Derrida's approach to *nuclear humanities* was heavily influenced by his concept of deconstruction, which emphasizes the ways in which the meaning of a text or concept is shaped by the context in which it is used. Derrida argued that the language and discourse surrounding nuclear weapons were saturated with contradiction and ambiguity, as political actors used rhetoric to justify their possession and potential use of nuclear weapons while also condemning their development and deployment by others. In his work, Derrida was particularly concerned with the way in which nuclear weapons had become a central part of the global political imagination, shaping not just the geopolitical landscape but also cultural and social practices. He argued that nuclear weapons represented a radical break from previous forms of warfare, and that their development had fundamentally altered the way that humanity thought about violence and power (Derrida 387). Derrida's work in *nuclear humanities* helped to establish a critical framework for understanding the cultural, political, and ethical implications of nuclear weapons. By focusing on the contradictions and ambiguities inherent in the language and discourse surrounding nuclear weapons, Derrida challenged scholars and policymakers to think deeply about the way that these weapons had shaped the global landscape, and to consider the ethical imperatives that lay at the heart of debates surrounding nuclear arms control and disarmament, launched by anti-war and anti-nuclear protests (Badruddin 8).

The Chernobyl disaster in 1986 represented another significant turning point for *nuclear humanities*. The nuclear catastrophe highlighted the risks and hazards of nuclear technology, and its aftermath prompted widespread discussion and debate about the role and responsibility of nuclear power (Plokhy 8). One of the most significant contributions to *nuclear humanities* after Chernobyl was made by the sociologist Ulrich Beck, who introduced the concept of *risk society*. In his book *Risk Society: Towards a New Modernity* (1986), Beck argued that modern societies had become increasingly dependent on technologies that posed significant risks and hazards, with the potential to cause catastrophic harm to individuals, communities, and the environment: '*The harmful, threatening, inimical lies in wait everywhere, but whether it is inimical or friendly is beyond one's own power of judgment, is reserved for the assumptions, methods and controversies of external knowledge producers*' (Beck 53–54). Beck argued that the Chernobyl disaster was an early warning sign of the dangers posed by modern industrial society and highlighted the need for greater accountability, transparency, and democratic oversight of technological developments (Beck 7). The cultural impact of Chernobyl was also significant, with many writers, filmmakers, and artists taking up the theme of nuclear technology and its risks, bright examples of which can be the Novel Award honoured writer Svetlana Alexievich's *Voices from Chernobyl* (1997) (Alexievich 2005) and the British novelist Jim Crace's *The Gift of Stones* (1988) (Crace 15), offering a haunting reflection on the fragility of civilization and the destructive potential of technology. These examples show that after the Chernobyl disaster *nuclear humanities* continued to engage with critical questions about the nature of humanity, technology, and risk, exploring the complex interactions between science, politics, and culture in shaping the nuclear age, or the *Nuclear Anthropocene*.

The Fukushima disaster (2011) was another significant event that impacted the field of *nuclear humanities*. The accident at the Fukushima Daiichi nuclear power plant in Japan resulted in the release of radioactive materials into the air and water, causing widespread devastation and prompting renewed scrutiny of the safety and ethics of nuclear technology (Juraku and Sugawara 1424). One of the most significant responses to the Fukushima disaster was the emergence of a new field of study known as *nuclear ethics*. This field explores the ethical, social, and political dimensions of nuclear technology, with a focus on questions of responsibility, accountability, and justice. Nuclear ethics scholars draw on a range of disciplines, including philosophy, political science, and environmental studies, to examine various aspects of nuclear technology, such as the safety of nuclear facilities, nuclear waste storage, and disarmament (Taebi and Roeser 5). Schools of thought in *nuclear humanities* after the Fukushima disaster continue to hold discussions and conferences to enhance the debates on the safety and ethics of nuclear technologies.

DEBATING THE CURRENT AGENDA OF *NUCLEAR HUMANITIES*

The agenda of *nuclear humanities* today reflects a growing concern about the risks and challenges, associated with nuclear technology, particularly as countries around the world are looking to expand their nuclear capabilities. (Verma iii-iv) by outlining the cultural and social perspectives on nuclear infrastructure, nuclear accidents, nuclear alternatives to fossil fuels to mitigate climate change, the global politics of nuclear weapons, including issues of nuclear deterrence, disarmament, and non-proliferation, as well as the ethical implications of nuclear waste disposal. Such agenda of *nuclear humanities* today is to deepen our understanding of the cultural, ethical, and environmental implications of nuclear technology and to foster informed and democratic debates about nuclear issues.

While having the nuclear as a societal debatable issue—"the good, the bad, the debatable" (Aref 9), *nuclear humanities* respond to these debating issues and involve hot discussions, revolving around ethical questions related to the use of nuclear technology, such as the morality of using nuclear energy as a power source or the decision to develop nuclear weapons (Friederich and Boudry) and the need for sustainable energy policies (Krupnik et al.). Other debates centre around cultural and social issues related to nuclear technology, such as the ways in which nuclear technology has been represented and understood in popular culture (Gorski) or the impact of nuclear technology on the way we think about risk and uncertainty (Kessides 3850). One ongoing debate within *nuclear humanities* is the question of how to balance cultural and political perspectives in the study of nuclear technology. Some scholars argue that a purely scientific approach to nuclear technology is insufficient and that cultural and social factors must also be taken into account (Downey 390; Alexis-Martin and Davies 3). Others believe that a focus on cultural and social factors can obscure the technical aspects of nuclear technology and hinder efforts to address practical issues related to nuclear energy and weapons (Budnitz and Rogner 540; Wilpert 4). Another debate within *nuclear humanities* revolves around the impact of nuclear technology on the environment and human health. (Ramana 127; Ekmekcioglu 2679) These debates within *nuclear humanities* are complex and multifaceted, reflecting the many different perspectives and concerns that surround nuclear technology.

Nuclear humanities is regarded to be a broader term that encompasses all aspects of human interactions with nuclear technology, from nuclear power plants to nuclear weapons, nuclear waste, and radiation, where the main concern is about the social, cultural, ethical, and political dimensions of nuclear technology, exploring how it shapes our society and how the society shapes it (Kaijser et al. 28). Among the issues, which enhance the humanities' perspective on the impact of nuclear technology on humanity, is the reflections about the role of nuclear energy in a nuclear war, "whose essential feature is that of being fabulously textual. . . . A nuclear war has not taken place: one can only talk and write about it . . . it is a non-event" (Derrida 23). Derrida focused on the ethical implications of nuclear weapons, arguing that their very existence raised profound questions about the nature of justice, sovereignty, and human responsibility. In his view, the sheer destructive potential of nuclear weapons made them a symbol of the fragility and vulnerability of human existence and underscored the urgent need for political action to prevent their use.

The use of nuclear technologies for producing nuclear weapons, with the potential of causing widespread destruction and catastrophic consequences, raises profound moral and ethical dilemmas, and questions the value of human life. In this context, *nuclear humanities* enhances the issues of researching the historical, cultural, philosophical, and ethical aspects of nuclear war via "framing a political consensus on the unacceptability of nuclear weapons" and "underlying moral ambivalence on the matter" (Hayashi 2) as well as provide a deeper understanding of its implications.

While communicating the ethical dilemmas that arise from the use of nuclear weapons, *nuclear humanities* analyses the moral implications of the use of nuclear weapons in warfare, such as whether it violates international law, is justifiable in certain situations, or undermines the moral principles that govern war:

> Using various philosophical concepts, it will explore the fundamental question as to whether any implementation of nuclear deterrence that involves a risk to civilians is morally acceptable. The models, though differing in origin and rationale, provide a unique lens from which to view this ethical dilemma. (Giampaolo)

By studying the cultural and psychological dimensions of nuclear war, *nuclear humanities* contributes to informing how societies can better prepare and cope with the aftermath of such an event or *nuclear trauma*, examining the psychological impacts of nuclear war on individuals, communities, and societies. By examining the cultural, social, and historical factors that lead to nuclear conflict, it is *nuclear humanities* that addresses the root causes of nuclear war and develops theories and solutions to prevent such conflicts from happening in the future, as well as deepening the understanding of how to promote peace, diplomacy, and ethical decision-making within the global community with the aim of shaping policies that prioritize peace, social justice, and human well-being. The literary and cultural perspectives on nuclear war, nuclear weapons, atomic bombing as well as various aspects of nuclear energy related issues, engaging with the nuclear, are the subject of study for nuclear criticism (Ruthven 10), examining how the nuclear has influenced and been represented in literature, film, art and other cultural forms, and how these representations intersect with larger socio-political concerns (Zapf 51). This field also aims to understand how nuclear power has shaped contemporary cultural values and attitudes towards nature, technology, and humanity.

The use of nuclear technology brings complex social and ethical issues that have far-reaching consequences. One key ethical concern in *nuclear humanities* is the question of whether nuclear technology should be used in the first place. *Nuclear humanities* can examine the ethical implications of nuclear energy, such as the risks of accidents, the proliferation of nuclear weapons, and the environmental impact of nuclear waste, as well as the ethical perspectives can be applied to issues like whether it is ethical to prioritize economic development over the potential dangers of nuclear energy. Another ethical consideration in *nuclear humanities* is the duty of nuclear powers to disarm and reduce the potential for a nuclear war, declared by the Science and Security Board of the Bulletin of the Atomic Scientists in their "Doomsday Clock Statement 2023." (Doomsday Clock Statement 2023), and making nuclear disarmament a topic of great importance in *nuclear humanities*, with scholars exploring issues like the efficacy of disarmament treaties and the role of civil society in pressing for nuclear disarmament, and other initiatives, trying "to place the disproportionate humanitarian consequences of nuclear weapon use at the centre of disarmament dialogue; and create a new international norm which would eventually have to be acknowledged" (Ritchie and

Kmentt 73). By considering the ethical implications of nuclear technology and weapons, scholars in *nuclear humanities* can help ensure that the future of nuclear technology is shaped by ethical considerations that prioritize the well-being of humans and the environment.

CONCLUSION

The interdisciplinary study of nuclear technology and its impact on society has been explored by scholars in various disciplines such as history, anthropology, literature, and the arts for many years. However, it was not until the twenty-first century that *nuclear humanities* became a recognized field of study, with the establishment of academic programs and research centres devoted to the topic. For example, Arizona State University and UT Austin offer specific degrees and courses in *nuclear humanities*. While remaining a growing area of research and advocacy, *nuclear humanities* also provide a critical lens through which to understand the broader cultural and social implications of nuclear energy and nuclear weapons. For example, the racial and economic disparities in the distribution of nuclear hazards, the complex global politics of nuclear proliferation, and the impacts of nuclear accidents (such as the Chernobyl disaster and the Fukushima catastrophe on local communities) are all complex problems that require a nuanced and interdisciplinary understanding, which *nuclear humanities* can provide. The humanities' perspectives provide a crucial space for critical reflection and inquiry into the complex and multifaceted issues surrounding the nuclear age and are essential for constructing a more comprehensive and ethical approach to our shared nuclear future.

In the future, *nuclear humanities* is likely to become an increasingly important field of study as the world continues to grapple with the challenges posed by nuclear technology. As the threat of nuclear war remains (United Nations Security Council Meetings Coverage), and as the dangers of climate change increase, there will be a growing need to understand the social, cultural, and ethical dimensions of nuclear technology and its impact on society. One area of focus for *nuclear humanities* in the future may be the development of new nuclear technologies, such as small modular reactors, that are seen as a more sustainable and environmentally friendly alternative to traditional nuclear power plants (Nakhle) with the focus on exploring the social and cultural implications of these new technologies, as well as the ethical and political questions they raise. Other areas of *nuclear humanities*' foci cover the issues of nuclear waste disposal, uranium extraction, advanced nuclear technologies, cyber security of nuclear safety application, etc., which requires a growing need for scholars in *nuclear humanities* to explore the cultural, social, and political dimensions of communicating the nuclear. *Nuclear humanities* is likely to remain among the important fields of study in the years to come, as the world continues to grapple with the complex and multifaceted challenges posed by nuclear technology.

The public debates about nuclear energy are complex and multifaceted, and there are valid for/against arguments on both sides: a central tenet of much of the pro-nuclear rhetoric is a misleadingly gloomy portrayal of renewable energy options, meanwhile, absolutist arguments against nuclear energy too often apply primarily to older plants no longer being built. And at times both sides tend to hang their hats on optimistic advances in technologies that may or may not become commercially available in time to make needed progress toward a low-carbon future, with references to such points as safety, cost, environmental impact, and energy security, when making decisions about the use of nuclear energy. Being one of agencies of ecological interactions among organisms, populations, communities and environments, nuclear energy development is an arena where ecological, political, and socioeconomic values collide at the debates, requiring critical ecological concepts. And *nuclear humanities* suggests such multidisciplinary tools addressing the challenges of the relationship between nuclear energy, ecosystems, environment, sustainable development, etc., in our energy-driven society.

REFERENCES

Alexievich, Svetlana. *Voices From Chernobyl: The Oral History of a Nuclear Disaster*, trans. Keith Gessen. Dalkey Archive Press, 2005.

Alexis-Martin, Becky, and Davies, Thom. "Towards Nuclear Geography: Zones, Bodies, and Communities." *Geography Compass*, vol. 11, no. 9, September 2017. https://doi.org/10.1111/gec3.12325

Aref, Lana. *Nuclear Energy: The Good, the Bad, and the Debatable. Learn More About Nuclear Technology, Its Benefits, and Its Dangers*. Massachusetts Institute of Technology, 2019. https://www.niehs.nih.gov/health/assets/docs_f_o/nuclear_energy_the_good_the_bad_and_the_debatable_508.pdf

Arendt, Hannah. *The Origins of Totalitarianism*. Cleveland: World Publishing Company, 1962.

Badruddin, Gowani R. *Global Peace and Anti-Nuclear Movements*. Mittal Publications, 2003.

Beck, Ulrich. *Risk Society: Towards a New Modernity*. SAGE Publications Ltd, 1992.

Budnitz, Robert J., Rogner, Holger H., and Shihab-Eldin, Adnan. "Expansion of Nuclear Power Technology to New Countries—SMRs, Safety Culture Issues, and the Need for an Improved International Safety Regime." *Energy Policy*, vol. 119, August 2018, pp. 535–544. https://doi.org/10.1016/j.enpol.2018.04.051

Campos, Luis. "The Birth of Living Radium." *Representations*, vol. 97, no. 1, 2007, pp. 1–27. https://doi.org/10.1525/rep.2007.97.1.1.

Carpenter, Ele. "Shifting the Nuclear Imaginary: Art and the Flight from Nuclear Modernity." *Cold War Legacies: Systems, Theories, Aesthetics*. Ed. John Beck, Ryan Bishop. Edinburgh University Press, 2016, pp. 116–133.

Gorski, Chris. "How Culture Wrestled with the Atomic Age." *Inside Science*, Wednesday, July 15, 2021. https://www.insidescience.org/manhattan-project-legacy/atomic-pop-culture

Crace, Jim. *The Gift of Stones*. Penguin Books, 2003.

Derrida, Jacques. "No Apocalypse, Not Now (Full Speed Ahead, Seven Missiles, Seven Missives," trans. Catherine Porter and Philip Lewis. *Diacritics*, vol. 14, issue 2, 1984, pp. 20–31. https://doi.org/10.2307/464756

Doomsday Clock Statement 2023. "It is 90 Seconds to Midnight." *Bulletin of the Atomic Scientists*. January 24, 2023. https://thebulletin.org/doomsday-clock/current-time/

Downey, Gary L. "Risk in Culture: The American Conflict over Nuclear Power." *Cultural Anthropology*, vol. 1, no. 4, 1986, pp. 388–412. http://www.jstor.org/stable/656378

Ekmekcioglu, Ozgul, Evangelista, Laura, and Kunikowska, Jolanta. "Women in Nuclear Medicine." *European Journal of Nuclear Medicine and Molecular Imaging*, vol. 48, 2021, pp. 2678–2679. https://doi.org/10.1007/s00259-021-05418-9

Feldman, Brian. "After the Catastrophe: Working with the Intergenerational Transmission of Collective Trauma in Jungian Analysis." *The Journal of Analytical Psychology*, vol. 67, no. 1, February 2022, pp. 105–118. https://doi.org/10.1111/1468-5922.12752

Friederich, Simon, and Boudry, Maarten. "Ethics of Nuclear Energy in Times of Climate Change: Escaping the Collective Action Problem." *Philosophy & Technology*, vol. 35, article 30, 2022. https://doi.org/10.1007/s13347-022-00527-1

Gager, Charles Stuart. "Some Physiological Effects of Radium Rays." *The American Naturalist*, vol. 42, no. 504, 1908, pp. 761–778. http://www.jstor.org/stable/2455770

Giampaolo, Kayla. "Deterrence or Disarmament?: The Ethics of Nuclear Warfare." *Carnegie Council for Ethics in International Affairs*, February 4, 2016. https://www.carnegiecouncil.org/media/article/deterrence-or-disarmament-the-ethics-of-nuclear-warfare

Hayashi, Nobuo. "On the Ethics of Nuclear Weapon. Framing a Political Consensus on the Unacceptability of Nuclear Weapons." *ILPI-UNIDIR NPT Review Conference*, series no. NPT2016, pp. 1–8. https://www.files.ethz.ch/isn/190435/on-the-ethics-of-nuclear-weapons-en-627.pdf

Juraku, Kohta, and Sugawara, Shin-Etsu. "Structural Ignorance of Expertise in Nuclear Safety Controversies: Case Analysis of Post-Fukushima Japan." *Nuclear Technology*, no. 207 (9), 2021, pp. 1423–1441, doi: 10.1080/00295450.2021.1908075

Kaijser, Arne, Josephson, Paul, Meyer, and Jan-Henrik. "Nuclear Society Relations from the Dawn of the Nuclear Age." *Engaging the Atom: The History of Nuclear*

Energy and Society in Europe from the 1950s to the Present, ed. Arne Kaijser, Markku Lehtonen, and Jan-Henrik Meyer. West Virginia University Press, 2021, pp. 27–51.

Kessides, Ioannis. "Nuclear Power: Understanding the Economic Risks and Uncertainties." *Energy Policy*, vol. 38, no. 8, 2010, pp. 3849–3864. doi:10.1016/j.enpol.2010.03.005

Krupnik, Seweryn, Wagner, Aleksandra, Vincent, Olga. "Beyond Technology: A Research Agenda for Social Sciences and Humanities Research on Renewable Energy in Europe Energy." *Research & Social Science*, vol. 89, no. 102536, July 2022. doi: 10.1016/j.erss.2022.102536, https://www.sciencedirect.com/science/article/abs/pii/S2214629622000433

Kuhn, Thomas S. *The Structure of Scientific Revolutions*, edition 2nd., Chicago: University of Chicago Press, 1970. https://www.lri.fr/~mbl/Stanford/CS477/papers/Kuhn-SSR-2ndEd.pdf

Malley, Marjorie. "The Discovery of Atomic Transmutation: Scientific Styles and Philosophies in France and Britain." *Isis*, vol. 70, no. 2, 1979, pp. 213–223. http://www.jstor.org/stable/230788

Nakhle, Carole. "The Potential of Small Nuclear Reactors." Geopolitical Intelligence Services. GIS Reports Online, March 17, 2022. https://www.gisreportsonline.com/r/small-modular-reactors/

Nuclear Energy and Sustainable Development. *World Nuclear Association*. October 2022. https://world-nuclear.org/information-library/energy-and-the-environment/nuclear-energy-and-sustainable-development.aspx

Plokhy, Serhii. *Chernobyl: The History of a Nuclear Catastrophe*. Basic Books, 2018.

Radioactive Waste Management. World Nuclear Association, January, 2022. https://world-nuclear.org/information-library/nuclear-fuel-cycle/nuclear-wastes/radioactive-waste-management.aspx

Ramana, M.V. "Nuclear Power: Economic, Safety, Health, and Environmental Issues of Near-Term Technologies." *Annual Review of Environment and Resources*, vol. 34, no. 1, 2009, pp. 127–152. https://doi.org/10.1146/annurev.environ.033108.092057

Rhodes, Richard. *The Making of the Atomic Bomb*. Simon & Schuster: Anniversary, 1986.

Ritchie, Nick, and Kmentt, Alexander. "Universalising the TPNW: Challenges and Opportunities." *Journal for Peace and Nuclear Disarmament*, vol. 4, no. 1, 2021, pp. 70–93.

Russell, Bertrand. *The Basic Writings of Bertrand Russell*. Edited by Robert E. Egner and Lester E. Denonn. With an introduction by John G. Slater. Routledge Classics, 2009.

Ruthven, Ken. *Nuclear Criticism: Interpretations Series*. Melbourne University Press, 1989.

The Ethics of Nuclear Energy: Risk, Justice, and Democracy in the Post-Fukushima Era. Ed. Behnam Taebi and Sabine Roeser. Cambridge University Press, 2018.

United Nation. Security Council Meetings Coverage. 9300th Meeting. SC/1525031, March 2023. https://press.un.org/en/2023/sc15250.doc.htm

Verma, Aditi. "The Nuclear, Humanities, and Social Science Nexus: Challenges and Opportunities for Speaking Across the Disciplinary Divides," *Nuclear Technology*, vol. 207, no. 9, 2021, pp. iii–xv. doi: 10.1080/00295450.2021.1941663

Wilpert, Bernhard. "The Relevance of Safety Culture for Nuclear Power Operations." *Safety Culture in Nuclear Power Operations*, ed. Wilpert, B., and Itoigawa, N. CRC Press, 2001, pp. 4–18. https://doi.org/10.1201/9780203302156

Zapf, Hubert. "Politicized Ecocriticism: From Nature Worship to Civilizational Critique." *Literature as Cultural Ecology: Sustainable Texts*, Environmental Cultures Bloomsbury Collections. Bloomsbury Academic, 2016, pp. 51–60. http://dx.doi.org/10.5040/9781474274685.ch-007

Chapter 19

Critical Animal Studies

Toward the Anthropocene

Ziba Rashidian

Critical Animal Studies, it could be argued, is less a sub-category of ecocriticism, than its twin. Both fields come into existence in the 1970s, each challenging in its own way the dominant assumption of human exceptionalism (human beings are not subject to nature and we can fix whatever we break!) and anthropocentrism (nature, nonhuman animals, and the planet are here for our use and convenience). In the over 40 years of their co-existence, both fields have registered seismic shifts in the way they conceptualize their areas of concern, given the unignorable environmental devastation that has occurred globally and the recognition by scientists of the sixth great extinction event. Thus, we see a convergence between the two fields such that CAS can be seen as an eco-concept contributing to the widening of the ecocritical conversation. CAS brings some specific arguments and lines of inquiry to the table. It has consistently and successfully challenged anthropocentrism, clearing the way for scholars to explore the moral, political and philosophical meaning of nonhuman animal lives and the way our human lives are inextricably bound up with what scholar Thom van Dooren calls "the more than human world." In challenging anthropocentrism, CAS has called into question the foundation of the Humanities, since they derive from what scholars have shown is a less-than-innocent drawing of a line between the human and all other forms of life. CAS scholars have gradually dismantled all the ways in which that line has been maintained, such as claims that only humans have culture, language, tool-making abilities, emotions, or minds, for example. As we move deeper into the *Anthropocene* and the drastic changes it portends, CAS investigates both the costs of our convenience and/or indifference to nonhuman lives from farmed animals to laboratory animals, but also strives, as a political and moral matter, to make us aware of our interdependence with the other creatures with whom we share the planet, the facts of our co-evolution with them, and the way that co-evolution points to our kinship with these other beings whose deep histories intersect with our own.

CAS, while it has antecedents in the anti-cruelty and vegetarian movements of the nineteenth century, takes its contemporary form thanks to two twentieth century moments of recognition: the first is the foundation of the environmental movement with writings by figures like Rachel Carson who registered the impacts on wild living nonhuman animals of our transformations of their habitats and the introduction of toxins like DDT into the biosphere. The second moment emerges with the advent of the factory farm or Concentrated Animal Feeding Operations (CAFOs) and industrialized slaughter which provided the impetus for the animal rights/animal liberation movement. The field has moved from discussions of animal rights and the possibility of providing legal protections beyond welfarist measures to analyses of the biopolitical or the political "set up" and ontological premises that enable our exploitation of nonhuman animals whether they are farmed animals, experimental subjects in laboratory settings, or *pets* kept for our convenience, status, or pleasure. As geologists move closer to labeling our present geological era, the *Anthropocene*, CAS has taken up the problems of climate change, habitat loss, and extinction as scholars strive to articulate our ethical and moral obligations to "free" living nonhuman animals. Scholars ask, what does it mean for one species (*homo sapiens*) to become a "geologic force" capable of transforming the planet so intensively that it threatens the continued existence of all the other life forms that came into being and thrived alongside what is sometimes called the "Cenozoic achievement"? In what follows, I chart three key inflections in the development of the field—the discussion of animal rights, the continuing examination of our relationship to nonhuman animals as a function of biopolitics, and the challenges faced by CAS within the context of the sixth mass extinction and the *Anthropocene*. Finally, in examining these ongoing conversations, I trace out the way sentience, or its various stand-ins, for instance, the capacity of "response" or for "agency" and/or "resistance," is decentered in the discourses emerging from *Anthropocene* and extinction scholarship.

The publication in 1975 of Peter Singer's *Animal Liberation: A New Ethics for our Treatment of Animals* represents an important moment in the development of CAS and sounds the call to action which is foundational for the field: he calls for an end to the "tyranny of the human over nonhuman animals. This tyranny has caused and today is still causing an amount of pain and suffering that can only be compared with that which resulted from the centuries of tyranny by white humans over black humans. The struggle against this tyranny is a struggle as important as any of the moral and social issues that have been fought over in recent years" (Singer 21). Singer's *Animal Liberation* articulates two key areas of investigation that become central for CAS: *speciesism* and *sentience.* Speciesism, a term first coined by Richard Ryder in the 1970s, identifies a type of species prejudice that Ryder thought paralleled other forms of discrimination such as racism and sexism. Singer picks up the term which, due to the success of his *Animal Liberation,* becomes widely known and used. Singer defines speciesism as "a prejudice or attitude of bias in favor of the interests of members of one's own species and against those of members of other species. . . . If possessing a higher degree of intelligence does not entitle one human to use another for his or her own ends, how can it entitle humans to exploit nonhumans for the same purpose?" (Singer 33). At its inception, speciesism was a conceptual tool used to challenge common assumptions undergirding the separation between human beings and nonhuman animals and providing justification for human use and dominion over nonhuman animals. The term itself challenges a set of ideas that affirm human exceptionalism, placing human beings outside of nature, and defining the human as something other and distinct from the animal. This is why Ryder began the practice of referring to "animals" as nonhuman animals—to reinforce the understanding of human beings as members of a species, *homo sapiens*, and thus as a kind of animal among other animals. Speciesism, likewise, highlights the fact that human beings are a species among other species and points to Darwin's claim that the differences in mind, are differences of degree, not differences of kind. Thus, the line of analysis that emerges from Singer's insights is that, for example, nonhuman animal sentience, emotion, and intelligence are not somehow wholly different from these capabilities and qualities in human beings.

The second area of investigation raised by Singer that is crucial for CAS is the moral significance of sentience and the denial or attribution of sentience to nonhuman animals impacted by human action and use. Singer's *Animal Liberation* scrutinizes animal experimentation, hunting, and our use of nonhuman animals for food. He is clearly responding to the emergence of factory farms (the first chapter is titled "Down on the Factory Farm"), or what are now called CAFOs. He draws on utilitarian thought to examine the treatment of animal lives in these operations, citing Jeremy Bentham's response to those who argue that animals make no moral claims on us because they are not rational beings, or lack other characteristics that we see as distinctly morally significant; Bentham's oft-quoted statement is: "The question is not, Can they reason? nor, Can they talk? but, Can they suffer?" (Singer 33). Singer's formulation of the moral significance of sentience (the ability to experience pain and enjoyment) is foundational for the field. Singer claims: "If a being suffers there can be no moral justification for refusing to take that suffering into consideration. No matter what the nature of the being, the principle of equality requires that its suffering be counted equally with the like suffering. . . . If a being is not capable of suffering, or of experiencing enjoyment or happiness, there is nothing to be taken into account. . . . So the limit of sentience . . . is the only defensible boundary of concern for the interests of others" (35).

Sentience has become a crucial area of scholarly investigation and sometimes debate in CAS with the majority of scholars making it central to their understanding of the ethical and political stakes of the field while others argue for a more biocentric approach—suggesting that sentience may construct yet another hierarchy in a field that has worked diligently to challenge such constructs. At present, however, sentience has proven to be a durable touchstone for CAS. It is sentience, rather than language, culture, tool use, social organization, or the mere fact of "being alive" that makes a living being "morally considerable." As Gary Varner outlines in his discussion of the term, "sentience" in our common understanding can mean "the individual is conscious, intelligent, self-aware, having freedom of choice or autonomy, and qualifies for personhood" (358). He notes that in CAS, the term usually means that the individual is "capable of conscious suffering and enjoyment" an understanding that goes back to Singer's *Animal Liberation.* We can see the importance of this claim and how it shapes the advocacy and activism of animal studies scholars by looking at abolitionist proponent Gary Francione and Animal Ethics founder Oscar Horta. In an August 20, 2012 blog, Francione argues against those who might want to link the moral status of nonhuman animals to certain cognitive capacities beyond sentience. He says, "It's not a matter of how 'smart' animals are or whether they have mental capabilities that we recognize as being like ours. If they are sentient, that is the only characteristic that they need for us to have a moral duty not to use them as our resources."

(Abolitionist). Francione makes this point because as Varner notes, moral considerability can be trumped by moral significance, where the argument is made that the needs of an individual organism with higher cognitive or other qualities (such as intelligence, autonomy, personhood, etc.) might outweigh the ethical claim for moral consideration of a "merely" sentient organism. Such qualities are often those that most resemble human qualities or capabilities and thus, the more an organism resembles us, the greater its moral significance. This is the criticism that was ultimately leveled at the Great Ape Project, a campaign to work toward the legal recognition of higher primate as "persons," a move that would presumably extend all rights and protections of personhood to them. The problem for many scholars with GAP was that the argument in favor of personhood for primates was conditioned by the perception of primates as being more "like us," that is, more like *homo sapiens* in their cognitive, emotional, and ethological qualities.

For activist scholars like Francione and Horta, creating hierarchies based on categories beyond sentience mutes the ethical claim inherent in sentience and reflects a form of speciesism, since in this line of reasoning human needs will always have greater moral significance than those of any nonhuman animal. The Animal Ethics organization asserts the same position that sentience is "sufficient and necessary" to grant an individual organism ethical significance: "being sentient and not merely being alive is what matters . . . we should be concerned only about sentient beings" (animal-ethics). Francione argues that the focus on sentience does not create a hierarchy placing the sentient over the non-sentient "because sentience is a necessary as well as sufficient characteristic for a being to have interests in the first place" (Abolitionist). This assertion that sentience is the necessary and sufficient ground for moral considerability has been labeled by detractors as *sentientism*, though adherents of this position remain steadfast in their argument for the signal importance of sentience as the foundation for rights and moral significance.

Just as Bentham's focus on the possible suffering caused by the human treatment of nonhuman animals has been carried forward into our current era, a competing line of reasoning about nonhuman animals emerges from the Cartesian tradition which sees animals as automata and interprets behavior as the result of reflex rather than consciousness or sentience. One could argue that the Cartesian argument lurks in the contemporary instrumentalization of animal bodies in CAFOs, since the animal caught within this "operation" is viewed only through the lens of maximal growth with the shortest amount of time to slaughter. Animals in these feeding operations have little to no possibilities of movement, socialization, or control over their use for reproduction. A good example of this efficiency can be seen in China's development of a new form of CAFO. A November 2022 headline in *The Guardian* announces "China's 26-storey pig skyscraper ready to slaughter 1 million pigs a year." The designer suggests the new type of facility will be more efficient and "intelligent" and involves both greater automation in the feeding of the pigs and resolves the problem of dealing with the waste produced by such a concentrated population by using the manure to produce biogas to run the facility (Guardian).

Of course, it's not just the lives of farmed animals in CAFOs that might raise questions about suffering; it is also the industrialization of their slaughter. Many investigators have published accounts of animals who are supposed to be stunned or rendered unconscious undergoing the processes of slaughter fully conscious. Dinesh Wadiwel gives such account in his *The War Against* Animals:

> An essential feature of contemporary industrialised chicken slaughter is the use of a "live hang." Chickens arrive at processing plants, packed tightly into crates. Workers open the crates, seizing the live chickens one by one by their legs, and hanging them upside down on fast moving conveyor hangers. The birds will proceed swiftly through the next stages of the mechanical process of their transformation from living being into dead meat. The birds will be led through an electrical water bath designed to stun them into senselessness, their necks cut, bled, and then their bodies scalded. (1)

He notes that because the assembly line process works so swiftly not every bird is unconscious as it goes through this conveyor belt of disassembly. Similar processes are in place for cattle, with a chain being attached to a leg and the steer being hoisted into the air, eventually having its throat cut. Horta describes cattle slaughter noting, "they are often not really stunned, and they are sometimes fully conscious when they are cut into pieces and skinned. Another worker in a slaughterhouse reported that he had seen many cases in which this happens. He said: 'the head moves, the eyes are wide and looking around. . . . They die piece by piece'" (Making a Stand for Animals 119).

To give a sense of the contemporary intensive use of animals for food worldwide, consider the following data. The website "Our World In Data" estimates the following numbers of animals slaughtered in 2022: a total of 76 billion animals were slaughtered, including 73.79 billion chickens, 1.4 billion pigs, 6.17 million sheep, 500 million goats, and 331.95 cattle. These numbers do not include the number of farmed and caught fish. Numbers of individual fish killed for the food industry are difficult to estimate since fish harvests are quantified as biomass tonnage rather than as individual animals (Mood et al.). Mood et al. estimate 124 billion fish were farmed in 2019 and that the majority of these fish are not protected by regulations on humane slaughter, suggesting they are not stunned before being processed. Both Faunalytics and Our World In Data project a continued intensification of the numbers of animals captured in what Wadiwel calls "the animal industrial complex" as the model of intensive farming of nonhuman animals developed in the industrialized West continues to be adopted globally.

While the focus on sentience has been particularly salient for CAS scholars focused on our intensified manipulation and exploitation of nonhuman animal lives in industrial agriculture, other scholars, confronting the impacts on animal lives due to climate change and other aspects of the *Anthropocene,* have shifted their focus to the significance and ethical questions raised by the what appears to be the sixth mass extinction event on the planet. Scientists studying the rate of extinction estimate that the current rate of extinction is between 1000 and 10,000 times the background extinction rate, or the rate at which species under "normal" circumstances go extinct, addressing extinction has become imperative in the field. CAS scholars who work on issues of the *Anthropocene* and extinction provide a different understanding of the relationship between the sentient individual member of a species and the species life that is lost in an extinction event. The analysis provided by these new scholars goes beyond what we might find in an animal ethics activist like Oscar Horta who argues that if we encounter a suffering wild animal, we should not abandon them and "let nature take its course," but rather as ethical beings we should respond to the animal's need and try to assist it. The new scholarship goes beyond this sentience-based, individualistic understanding of ethics and reframes the biocentrist argument that all life, sentient or not is valuable. An example of the biocentrist argument can be found in Stephanie Jenkins "Returning the Ethical and Political to Animal Studies," where she argues that "Until we recognize the lives of all animate beings as worth protecting, the hierarchical dualisms of human/animal, mind/body, and nature/culture will remain intact" (505).

The work of Thom Van Dooren exemplifies the way the new scholars focused on the *Anthropocene* and the Sixth Mass Extinction are attempting to move the conversation away from both sentientism and biocentrism. In *Flightways: Life and Loss at the Edge of Extinction*, Van Dooren asks us to rethink our understanding of extinction—both our idea that extinction occurs when the last individual representative of a particular species dies and that extinction, like the extinction event now unfolding, is something that happens "out there" in "nature" and has no direct implications for human beings, other than a diminishment of biodiversity as something valuable in itself. Van Dooren, through careful field work with endangered animals and through the creation of narratives about both the natural history of the species he is studying and the individuals he observes, allows us to see both the suffering of the individual animal and asks us to examine what is lost well before the last of a kind slips out of the world. Van Dooren notes that by the time we get to the last exemplar, extinction has already occurred because extinction is not simply about the loss of a particular kind of being (a species or life form); it is also about a particular "way of life," the complex worlding that a species co-produces with others of its kind across generations, and with the network of life that sustains its existence.

CAS puts the nonhuman animal at the center of its concerns, examining what enables our daily, casual use of their bodies as food, their lives as tools for our research, their merchandising as commodities as in the exotic pet trade, or their controlled and curtailed existences as our companions. The goal is to make visible what we either cannot see or would rather not see: the deformation of lives in CAFOs, the realities of slaughterhouses, or the deaths of what we categorize as wild animals resulting from climate change, environmental degradation, or the human appropriation of the livable earth.

Perhaps one of the most crucial debates percolating in the field at present is that between those who, like Horta and Francione, argue that sentience is absolutely essential as the foundation for animal ethics and those who, faced with what they see as the unraveling of the web of life, want to argue from a biocentric position that *life* itself is sufficient, pointing to all the non-sentient forms of life that are necessary for our own life and for the life of ecosystems. Critics of this position often dismiss it by asking whether an intestinal parasite or bacterium has "moral considerability" such that we pause before seeking medication to free ourselves from such an uninvited tenant.

A second issue confronting CAS lies in its monofocalism: the way, in general, it focuses either on farmed animals (as I have done here) or on *wild* animals, particularly endangered species. While, of course, many scholars do examine the issues around petkeeping, animal experimentation, and animals in entertainment, the present moment is marked by two unignorable real-world realities: the intensification of our industrialization of animal lives for food and the increasing evidence of the sixth mass extinction event on our planet. As David Wallace-Well's notes, "Ninety-six percent of the world's mammals, by weight, are now humans and their livestock; just four percent are wild" (Wells 154). Given the numbers, the sheer "weight" of those lives that count as *livestock*, it is clear why many CAS scholars focus on farmed animals. But the issue of wild animals and what constitutes the *wild* when there appears to be, as Bill McKibben claims in *The End of Nature,* no *nature* remaining that is free of the effects of modern industrial society is still a developing element of CAS discourse. Some recent scholars working on the wild animal question have also turned to Foucault and biopolitics to address the kinds of lives possible for wild or free-living animals. Ethicists, too, have struggled to bring the non-domesticated animal into the discussion of the moral obligations we have to nonhuman animals. Van Dooren sees our ethical entailment as involving both bearing witness to the lifeways and lives that are being lost in extinction and struggling to tell the stories of those who are being lost. Philosopher Oscar Horta, on the other hand, makes an argument for more direct forms of response, arguing that if we see an animal suffering in the wild, we should feel compelled to assist that animal and not revert to the idea that a wild animal's life is subject to the violence of nature's struggle for existence, a struggle that makes no moral claim on us. But if nature no longer exists as some sort of pristine realm beyond human intervention, then the suffering of the nonhuman animal makes an ethical claim on us (247ff).

A final issue facing CAS is the obstacle it has consistently challenged—the persistence of anthropocentrism. Despite decades of critiquing human exceptionalism, CAS appears to have made little headway when it comes to real-world conflicts between human and nonhuman animals. One can think of the persistent conflict between lumber industry demands to be able to harvest old growth forests and the often-threatened animals, like the spotted owl, that reside in those forests. For the lumberjacks the harvest means employment and security for their families; for the owls and other animals living within the old forest ecosystem, it means their death and destruction. Or, one can think of the persistent conflicts between farmers in Indonesia and Kenya and elephant populations resulting in the poisoning (in Indonesia) and the shooting (in Kenya) of so-called "problem" elephants, with the result that elephants are now endangered. Certainly, as climate change and population pressures continue to grow, such conflicts between the human and nonhuman animals over access to the livable earth will only intensify. Still less has CAS been able to conquer the anthropocentrism that shapes our day-to-day lives; anthropocentrism thrives in the casual decisions about our diets or in our attitudes to petkeeping; it is our human concern and convenience that shape our understandings and our choices. Even Wallace-Wells confesses, "I may be in the minority in feeling that the world could lose much of what we think of as 'nature,' as far as I cared, so long as we could go on living as we have" (36). In fact, later in the book, he'll equate the stories told about the impacts of climate change on *wild* animals as parables that succumb to the "pathetic fallacy" (150). Of course, his book is geared to prompt human beings to recognize the realities of climate change and to push for the large actors (governments, industry) to take action and so necessarily his argument appeals to our self-concern, our species-being, our anthropocentrism. Still, he is not alone in his focus on the effects of climate change on human populations and it is the relative silence on the nonhuman animal question that reminds us anthropocentrism is very much with us. And thus I would argue that CAS in its grappling with anthropocentrism illustrates a challenge that is fundamental to our articulation of eco-concepts suited to address the *Anthropocene* and whatever dispensation comes after.

REFERENCES

"Battles over Ever-Decreasing Land." World Wildlife Fund. https://wwf.panda.org/discover/knowledge_hub/endangered_species/elephants/human_elephant_conflict/. Accessed November 1, 2023.

"China's 26-Storey Pig Skyscraper Ready to Slaughter 1 Million Pigs a Year." *The Guardian*. Guardian News and Media, November 25, 2022. www.theguardian.com/environment/2022/nov/25/chinas-26-storey-pig-skyscraper-ready-to-produce-1-million-pigs-a-year.

Francione, Gary and Charlton, Anna. *Animal Rights: The Abolitionist Perspective*. Exempla Press, 2015.

Francione, Gary. "Only Sentience Matters." Animal Rights: The Abolitionist Approach. August 20, 2012, www.abolitionistapproach.com/only-sentience-matters/.

Horta, Oscar. *Making a Stand for Animals.* Routledge, 2022.

"How Many Species Are We Losing?" WWF, World Wildlife Fund, wwf.panda.org/discover/our_focus/biodiversity/biodiversity/. Accessed November 1, 2023.

Jenkins, Stephanie. "Returning the Ethical and the Political to Animal Studies." *Hypatia*, Summer 2012, vol. 27, no. 3, pp. 504–510.

McKibben, Bill. *The End of Nature*. Random House, 2006.

Mood, A., Lara, E., Boyland, N., and Brooke, P. "Estimating Global Numbers of Farmed Fishes Killed for Food Annually from 1990 to 2019." *Animal Welfare,* 32, E12. doi:10.1017/awf.2023.4

"Old-Growth Forest in Humboldt County Spared from Clearcutting." Center for Biodiversity. https://www.biologicaldiversity.org/news/press_releases/2013/old-growth-forest-01-23-2013.html. Accessed November 1, 2023.

Singer, Peter. "All Animals Are Equal." *Writings on an Ethical Life.* HarperCollins, 2000, 28–46.

Van Dooren, Thom. *Flightways: Life and Loss at the Edge of Extinction.* Columbia University Press, 2014.

Varner, Gary. "Sentience." *In Critical Terms for Animal Studies.* Lori Gruen (ed.). University of Chicago Press, 2018, 356–69.

Wadiwel. Dinesh. "Biopolitics." In *Critical Terms for Animal Studies.* Lori Gruen (ed.). University of Chicago Press, 2018, 79–98.

Wallace-Wells, David. *The Uninhabitable Earth: Life After Warming.* Tim Duggan Books, 2019.

Chapter 20

Unravelling Critical Plant Studies

Kübra Vural Özbey

Albeit being the first organisms on earth, plants have been long reduced to non-existence in the Western world centred by the hierarchical dichotomy between human and non-human realms. It is true that plants are considered useful as a source of nutrition, for medical practice, and in various industries for production. However, it would not be a mistake to state that the boundaries nascent in the Western chain of beings marginalise the plants at the bottom of the hierarchy with "the lowest limit of the living," place animals in a liminal position with their "some higher functions or ends," and put human beings on a pedestal (Nealon 32–33). This system of thought is amplified in philosophical, ethical, and scientific discourse and conventions to the extent that plants are regarded as immobile, insentient, and almost non-living objects. Yet, the ecocritical turn in humanities, encompassing the intersection of different fields, strives to challenge the well-established perception of the othered nature in Western thought. While ecocriticism reconsiders the nature-culture dichotomy, it prioritizes the ontological and epistemological dimensions of non-human beings by highlighting their agential power. To be more precise, the rhizomatic structure of ecocritical studies, referring to the interconnection of ecocriticism with other branches of movements and studies, operates to dissect the agency of non-human beings, the interconnection between humans and non-humans, and the intrinsic value of marginalized nature with its focus on new materialisms, transcorporeality, and posthumanism. In light of this new paradigm, there is an acknowledgement of the vegetal realm as part of non-human entities in a critical framework, investigating and questioning the passive role attributed to plants. What broadens the discussion on plants is Critical Plant Studies (CPS), also sometimes referred to as Human-Plant Studies (HPS), which emerges as an eco-concept embellished with the radical shift in perspective following the novel trends in ecocritical studies. This chapter sets out to ground Critical Plant Studies, grapple with the scientific and philosophical frameworks to comprehend plant agency and locate the issue in literary studies to unveil the future projections of this eco-concept in humanities.

As a starting point, a brief look at the emergence, development, and main tenets of CPS is significant. In the early 2010s, CPS becomes apparent in social sciences and humanities by "consider[ing] the histories and the power dynamics involved in our utilitarian relations with the vegetal world; borrowing from emerging insights in the Natural Sciences and from much older forms of plant knowledge; it also considers plants as living organisms with their own forms of agency, being, and desire" (Sandilands 157). The first thing one notices in CPS is the emphasis on plants' cognition, their way of communication, and intelligence agency. This approach yields a great challenge to the conventional understanding of plants without agential power. As an interdisciplinary area, CPS also brings a clearer focus on the vitality of plants akin to that of animals. While Western ethics and philosophy is grounded on zoocentrism, CPS inveighs against animal-centred onto-epistemological outlook. Matthew Hall casts light on this undertaking of CPS and states that "zoocentrism is a deliberate philosophical strategy for marginalizing and excluding plants. Zoocentrism is a *method* for achieving the exclusion of plants from relationships of moral consideration [and] helps to maintain human notions of superiority over the plant kingdom in order that plants may be dominated" (*Plants* 6). Accordingly, CPS unfolds the intrinsic value of the vegetal and aims to end the degraded position of plants and their sense of otherness compared to animals. Although there is a tendency to delineate CPS as "an extension of animal studies, similarly requiring a radical rethinking of the very nature of ontology and working toward a new understanding of vegetal subjectivity" (Vernon 93), the vegetal turn ostensibly springs from another impulse to overthrow the superior position of animals based on their so-called complex structure of being. Undoubtedly, both fields address the issue of otherness and marginalization of non-human beings, but CPS seeks to cross double boundaries to place the vegetal potential, action, and capacity without the web of service relationships to other non-human and human beings. This amounts to saying that CPS explores to establish the inherent value of the vegetal in new discursive practices. At stake here is to entail a discussion about the vibrant vegetal life from a plant-centred perspective, inaugurating a fresh dialogue about biodiversity. In this respect, CPS features a struggle against *plant blindness*, a term for "the literal ignorance of plants by human beings and their spontaneous preference for animal life" (Hall, *Plants* 5). More tellingly, James H. Wandersee and Elisabeth E. Schussler originally coined the phrase:

> (a) [T]he inability to see or notice the plants in one's environment; (b) the inability to recognize the importance of plants in the biosphere and in human affairs; (c) the inability to appreciate the aesthetic and unique biological features of the life forms that belong to the Plant Kingdom; and (d) the misguided anthropocentric ranking of plants as inferior to animals and thus, as unworthy of consideration. (82)

As noted above, the slighted treatment of plants in a human-centred paradigm can be unpacked in Wandersee and Schussler's plant blindness. Indeed, the phrase mirrors the problematic aspects that CPS dwells on in that the field of study endeavours to abandon the lack of connection between plants and humans and orients to appreciate the liveliness of plants. Taken as a whole, CPS maps out a trajectory to zoom in on the vegetal world through new lenses.

By dealing with CPS in more detail, it is necessary to frame the issue in relation to contemporary scientific research that advances the discussion in the field. To further understand the contribution of recent investigations to CPS, studies on neurobiology, plant intelligence, and communication, all of which are intertwined issues, can be elaborated here. To begin with neurobiology, the term signifies the illustration of plants' perception of the world, and their surroundings, and the identification of their network and nervous systems. Eric D. Brenner and et al. explain that it is essential to study plant signals to understand neurobiology, and the signals "include recent discoveries of intercellularly transported macromolecules that regulate development and/or defense pathways, including transcriptional activators, RNA molecules and peptide hormones, as well as decades-worth of information on phytohormones" (413). Although the neurobiological systems of humans and animals are already clarified in terms of their behavioural analysis, plant neurobiology is reconsidered in past decades thanks to some experiments and observations about their signalling methods and the change of phenotype. According to Brenner and et al.'s research, plants are capable of reacting to their environmental circumstances, and their complex signalling systems resemble the nervous systems of animals although theirs is of a different nature from having cellular structures or brains. This discovery paves the way for initiating an argument on plant intelligence considering that plants are recognized with their neurosensory systems now. The term, plant intelligence, does not rest on the premise that plants are intelligent in the sense that we define human intelligence. It rather alludes to plants' ability to adapt to their environments, to interact and interconnect, and their mobility. In stark contrast to the previous perception of plants as passive objects, their intelligence capacity is acknowledged as research unveils:

> [P]lants do in fact communicate with other plants; they evidence both defensive and aggressive behavior; they feign certain states to fool predators or attract pollinators; and of course plants do move, only at a much slower time-scale than most animals; there is even research to suggest that plants feel pain, or at least respond decisively to extreme danger. (Nealon 30)

More interestingly, the experiment known as *mimosa pudica* highlights that plants record and remember their own reactions at the moment of attack or danger to protect themselves (Rieger 59). Such studies serve the broad context of CPS on the ground that plants are proven to be vital agents, sensing, memorizing, acting, and transforming themselves. Lastly, the communication of plants is illustrative of the development of expressive forms in vegetal life. What interests us about vegetal communication is the use of language by plants. Monica Gagliano, John C. Ryan, and Patrícia Vieira examine the language of plants in two categories: extrinsic language and intrinsic language. While the first one refers to "the scientific language about plants (especially the taxonomic terminology that shapes how we speak and think about the botanical world), the philosophical language deployed to articulate the particularities of plant ontology, and the representation of vegetality in literary works" (Gagliano, Ryan and Vieira xvii), the latter does not incorporate a human-based approach to vegetal language. The scholars gloss intrinsic language as "the language of biochemistry—plant hormones, electrical signaling, pressure cues, and so on, as well as the multisensorial expressions of plants—their visual articulations, their olfactory bouquets, or their aural enunciations, revelated in the emergent field of plant bioacoustics" (Gagliano, Ryan and Vieira xviii). This type of language, hence, suggests a form of communication peculiar to plants, distinct from other human and non-human beings. The analysis of plant language is also linked with the study area of ecocritical biosemiotics in which the unique forms of vegetal expression are closely investigated in line with phytosemiotics which is the study of vegetal signs. The significance of such studies resides in their revelation of the onto-epistemological agency of the vegetal. The scientific enquiry of plant neurobiology, intelligence, and language unmasks the visibility of plant bodies as active agents and displays the knowledge-making process of plants. In this regard, the interdisciplinary ground of CPS becomes most evident. John Charles Ryan aptly puts:

> [E]xtending and building upon ideas of plant intelligence, learning, and behavior (as articulated in various areas of the biological sciences) to the humanities and social sciences, critical plant studies (CPS) attends to the life worlds of plants, including their predispositions for intention, sensing, sensation, meaning-making, memory, and forms of experience and expression. Through the integration of the humanities and sciences, plants can be understood more

> persuasively as agents acting upon—and, indeed, transforming—art, culture, literature, politics and all activities in which human beings engage. ("Planting" 64)

At this conjuncture emerges the interconnection between CPS and ecocriticism as well. Needless to say, ecocriticism specifies the relationship between human and non-human worlds by dwelling on the neglected existential and discursive metarealities of the natural realm. As CPS concentrates on plant awareness, intelligence, and language, its main objectives overlap with the tenets of ecocriticism so they augment a discussion on plant agency by presenting extensive examples of the way we should comprehend vegetal action and language.

Having established the contribution of scientific research to CPS, it is essential to lay out how Western philosophical thought on plants is challenged in this vegetal turn. As widely acknowledged, the Western mindset, shaped by Greek philosophical tradition at first, originally attributes passivity to plants. Starting from Plato's views, a short view of philosophical ideas sketches out the process that degrades the vegetal agency. According to Plato's writings, plants are deprived of reason and control as silent and inactive beings. In the tripartite structure of the soul, Plato argues that plants, slaves, and women occupy the lowest degree of "the appetitive soul" because they are thought to be inactive, irrational, and impulsive (Hall, *Plants* 20). Thus, Plato adopted an approach to the passivity of plants. Although Empedocles recognizes a relationship between plants and other beings before Plato, the descendants of Plato thoroughly dismiss an understanding of the vegetal world with its intrinsic value. As a case in point, the Aristotelian view of plants dominates and impinges on the standpoint of the vegetal. Structuring a strict hierarchical system of beings, Aristotle works on the classification of the soul into three groups: nutritive or vegetative soul, sensitive soul, and rational soul. Plants appear at the basic level of the vegetative soul as a reflection of the unreasonable side of the soul. What is peculiar to the vegetative soul is growth and reproduction. Jeffrey T. Nealon depicts this state as "sessile; it doesn't move but merely channels the principle of (uncontrolled) growth" (36), and plants fit into this category. Moreover, Aristotle contrasts plants with animals capable of action, perception, and sensitivity in the second form of the soul. At this point, it is necessary to keep in mind that Aristotle assesses plants in comparison with animals: "Aristotle almost always defines and evaluates plants using *zoological* criteria and considers plant physiology as a *series of lacks*" (Hall, *Plants* 27). Aristotle, therefore, adopts the zoocentric approach in the evaluation of plants, defining the future of Western philosophy. Despite the studies of Theophrastus rejecting zoocentrism in botanical research, the dualism that Aristotle offers has a great impact on medieval and early modern thinkers. While a discrepancy between plants-animals and plants-humans arises, Descartes's dualism then looms over the Western perspective fostering the idea of plant inferiority and passivity. In this background, however, popular radical ideas about the vegetal realm later on come from Erasmus Darwin and his grandson Charles Darwin. While the first acknowledges the mobility, sensitivity, and even intelligence of plants to a certain degree, (Gagliano, Ryan and Vieira xi), Charles Darwin, with the help of his son, mentions the plant brain (Barlow 37–38). The works of Darwin throw light on further research on plant agency that CPS leans on. In a similar fashion to ecocritical studies challenging the dualisms of nature-culture, mind-body, and human-non-human, CPS aims to dismantle the notions and practices of the Western mindset denying biodiversity and vegetal onto-epistemology.

Obviously, it is still demanding to go against the grain of Western philosophy on the matter of plant studies. CPS strives to make a paradigm shift in the evaluation of vegetal life by offering a plant-centred perspective. Accordingly, the discussion on CPS can be developed here to the construction of *plant personhood* and the subsequent questioning of plant ethics. With regard to plant personhood, it is important to note that it initially emanates from a non-Western worldview in which different groups of native people, associated with animist cultures, embrace a holistic attitude to nature. Graham Harvey, for instance, mentions the perspective of indigenous people and illustrates:

> Plants are engaged with as powerful persons, some of whom, like animals, give their lives or parts of themselves for the benefit of other persons (not only humans). Some, like tobacco, aid the rather feeble attempts of humans to communicate with the wider community of life, and especially with more powerful persons, by carrying prayers and invocations upward and outwards. That they join in ceremonies and cultural politics and hold ceremonies of their own is, for animists, demonstration enough that they are persons. (105)

Nurit Bird-David, too, refers to Nayaka, the community that develops a sense of kinship with nature by highlighting "we-ness" in their connection with other beings (73). Likewise, the Aboriginal people are known for their belief in their kinship with plants (Hall, "In Defence" 7). The idea that humans are connected with non-human beings in a network of kinship suggests the attribution of personhood to non-human beings. Thomas J. Puleo pinpoints six elements to distinguish personhood: "agency, sentience, perception, consciousness, communicability and contractibility, the last being the ability to enter into an agreement" (214). According to the experiments on plant agency, intelligence, and communication, vegetal beings can be defined in terms of personhood. Based on these aspects, however, this kind of extensive approach is a storm centre of many discussions in the Western tradition although animist groups of indigenous people have already adopted this idea. Acknowledging various arguments offered in this context, Puleo offers his own answer as follows:

> By creating different kinds of persons, based upon ontological differences, one can preserve distinctions among human, nonhuman animal, plant and mineral entities, yet recognise a basic commonality that can serve as a basis for respectful behaviour. In a way, the solution is similar to a legal fiction, only that it would be more rigorously informed by scientific research and humanistic thought. (226)

In a similar fashion, Matthew Hall reminds the flexible context of glossing personhood ("In Defence" 3) and highlights the importance of recognizing plant personhood since it allows us to hear the polyphonic "voices" of plants and get engaged with a lively conversation ("In Defence" 10). That is to say, the idea of plant personhood augmented by animist cultures leaks into the discourse and discussion of Western thought particularly after the rise of new plant studies.

Albeit the reception of plant personhood, much of the difficulty in ascribing personhood to plants in Western philosophy stems from ethical discussion. The radical shift in the appreciation of vegetal life leads scholars to reconsider our moral and ethical approach to plants. Ostensibly, moral questions on animal rights have already been met with critical acclaim in ethical discussions. Thus, extending the debate to plants makes a crucial point about an all-inclusive attitude towards the non-human world. Drawing on the idea of plant agency and personhood, scholars explore to find out the ways that plants can be treated as ethical subjects without being exploited as objects. In this regard, the incorporation of humans and plants is of significance in that moral sensibility can be established. It is known that ancient or indigenous societies, for instance, do not take violent actions against plants, as can be noticed in their cultural practices and texts (Hall, *Plants* 160). Fundamental to their moral responsibility is the strong bond of interconnection with the vegetal. The current studies on plants are at pains to revise this sensibility by providing an ethical ground for the care of non-human others. Although the discussion on plants marks a high point in theoretical criticism, the practice of moral responsibility in behavioural terms is still problematic. Anna Lawrence draws attention to the challenging ethical questions, stating: "This ethical rethinking collides with difficult questions of diet, climate change, race and capital amongst others—how can we best value plants within their entangled relations with other beings?" (9). Lawrence's probe into the matter might pose serious questions deep-seated in various studies, but it actually evokes the interdisciplinary range of debates in CPS considering that "a plant ethics cannot by its very nature be limited to the plants themselves, but is inherently multispecies, multi-scalar and 'multi-elemental'" in Lawrence's words (11). To put the matter simply, the practice of exploitation, slavery, and discrimination, whether it be to women, people of colour, animals or plants, is the main reason for moral problems. The sensible connection with one another, care, and respect affirm moral status in the relationship between humans and plants.

Also of importance is placing CPS in literary studies since the discussions of the vegetal turn take on diverse and enlarging meanings as exemplified in literary works. In a culture where people address the silent, immobile, and insensitive one as "vegetable" (James 254), it is crucial to remind the original roots of the word, hence the vegetal being, referring to liveliness, activeness, and growth (Marder 20). As with literary criticism and analysis, the task turns out to advance the argument of CPS in exemplary stories. One might propose that fictional narrations are rich in creating fantastical, not realistic, visions of plants about their agential potential. As a point of interest, Ryan gives instances of myths, fantasy literature, and children's literature ("Passive Flora" 102). On the other hand, the threads of literary works can be depicted as more considerate when it comes to the interconnection between human and non-human worlds. Fittingly, as is manifest in the works such as the medieval poem *The Dream of the Rood*, the Romantic poems like William Blake's "Ah, Sun-flower" and William Wordsworth's "To the Daisy," or a contemporary novel, *Garden by the Sea* by Mercé Rodoreda, the voice of the vegetal can be heard or centred more in the works of literature, granting a fertile ground to stress the behavioural and discursive potential of plants. More tellingly, CPS takes a keen interest in exploring "new possibilities for the stories we tell about plants" (Ryan, "Planting" 62). Suffice it to say, such "possibilities" can be vividly detected in various literary texts. As is evident in the integral role of literary criticism in ecocritical studies, the use of literature in CPS is essential, too, on the ground that it appeals to people more than scientific and philosophical debates. When the perspective CPS offers is applied to literary works, it rests on a distinction from conventional readings of vegetal symbolism and metaphors in such texts. It rather corresponds with the examination of texts in their gaps to reveal more about plant intelligence, language, personhood, and ethics. The insights that literature presents are rich mines for CPS as can be evidenced in a few examples here. *Plants and Literature: Essays in Critical Plant Studies* (edited by Randy Laist) examines the role of plants by emphasizing the interaction between the human and non-human world in a myriad of examples from the nineteenth century to contemporary works in the Western canon. *The Language of Plants: Science, Philosophy, Literature* includes a section in which two chapters propose the analysis of literary works, including an example from Turkish literature, at the intersection of ecocriticism and CPS. *Plants in Children's and Young Adult Literature* (edited by Melaine Duckworth and Lykke Guanio-Uluru) goes beyond the detection of plant symbolism in fantasy worlds and integrates an ecocritical approach to CPS to articulate new visions. Finally, the conference of Cappadocia University entitled *Critical Animal and Plant Studies* (May 16–18, 2022) puts forth fresh ideas in the evaluation of various literary works from different centuries in the new paradigm of plant studies. What matters more in these instances is that literary works abound in transferring the complex context of CPS directly in the exemplary cases they share so that the analysis of works in this light allows us to get the gist of the argument without difficulty.

Evidently, from the early 2010s to the first half of 2020s, CPS, as in the case of ecocriticism, has evolved in a rhizomatic structure with novel perspectives offered by science, philosophy, and literature. While scientific experiments on plants prove vegetal sentience, activity, and communication, philosophical debates have ample room for ethical and moral considerations in the approach to plants. Yet, compared to animal studies, there is still a gap to be addressed in CPS in terms of the ethical approach to plants. It is necessary to end the actions of exploitation and destruction of plants by intensifying moral responsibility and sensibility through recognition of plant agency. In this light, the connection between human and vegetal worlds has to be strongly established. Accordingly, the role of literary criticism and analysis must not be overlooked in the development of plant studies as its critical perspective operates to concretely illustrate vegetal agency to the reader. Moving on to the future of CPS, one may contend that the analysis of more literary texts from the lens of CPS will provide a rich context for social sciences and humanities. Moreover, it will enable to entail a paradigm shift in human recognition of vegetal turn, inducing possible behavioural change after raising awareness of plant cognition, personhood, and ethics. As a final note, the strength of CPS can be highlighted in its essence as an eco-concept. In effect, contemporary circumstances invite twenty-first-century people to take responsible actions at the turn of more ecological crises. The recognition of the intrinsic value of non-human worlds reinstates the loss of connection between humans and non-humans. In this vein, learning more about eco-concepts will assist to end the sense of otherness, and may provide new solutions to ecological predicaments. Accordingly, CPS tinged with ecocritical theory strives to solve the conflicts between humans and plants by establishing a dialogue based on appreciation, care, and respect.

REFERENCES

Barlow, Peter W. "Charles Darwin and the Plant Root Apex: Closing a Gap in Living Systems Theory as Applied to Plants." *Communication in Plants: Neuronal Aspects of Plant Life*, edited by F. Baluska, S. Mancuso, and D. Volkmann. Springer, 2006, pp. 37–51.

Bird-David, Nurit. "'Animism' Revisited: Personhood, Environment, and Relational Epistemology." *Current Anthropology*, vol. 40, no. 1, 1999, pp. 67–91.

Brenner, Eric D., et al. "Plant Neurobiology: An Integrated View of Plant Signaling." *Trends in Plant Science*, vol. 11, no. 8, 2006, pp. 413–419.

Critical Animal and Plant Studies. By Cappadocia University, May 16–18, 2022. https://ehc.kapadokya.edu.tr/MediaUploader/Documents/EH%20Conference%20Book%20of%20Abstracts_ds3.pdf.

Duckworth, Melanie, and Lykke Guanio-Uluru, editors. *Plants in Children's and Young Adult Literature*. Routledge, 2022.

Gagliano, Monica, John C. Ryan, and Patrícia Vieira. Introduction. *The Language of Plants: Science, Philosophy, Literature*, edited by Gagliano, Ryan, and Vieira. University of Minnesota Press, 2017, pp. vii–xxxiii.

Hall, Matthew. *Plants as Persons: A Botanical Philosophy.* SUNY Press, 2011.

———. "In Defence of Plant Personhood." *Religions*, vol. 10, no. 5, 2019, pp. 1–12. mdpi.com, https://doi.org/10.3390/rel10050317.

Harvey, Graham. *Animism: Respecting the Living World.* Columbia University Press, 2006.

James, Erin. "What the Plant Says: Plant Narrators and the Ecosocial Imaginary." *The Language of Plants: Science, Philosophy, Literature,* edited by Monica Gagliano, John C. Ryan, and Patrícia Vieira. University of Minnesota Press, 2017, pp. 253–272.

Laist, Randy, editor. *Plants and Literature: Essays in Critical Plant Studies*. Rodopi, 2013.

Lawrence, Anna. "Listening to Plants: Conversations between Critical Plant Studies and Geography." *Progress in Human Geography*, vol. 46, no. 16, December 2021, pp. 1–13. doi: 10.1177/03091325211062167.

Marder, Michael. *Plant-Thinking: A Philosophy of Vegetal Life.* Columbia University Press, 2013.

Nealon, Jeffrey T. *Plant Theory: Biopower & Vegetable Life*. Stanford University Press, 2016.

Pollan, Michael. "The Intelligent Plant." *New Yorker,* December 23, 2013, https://www.newyorker.com/magazine/2013/12/23/the-intelligent-plant.

Puleo, Thomas J. "Incorporating Nonhuman Subjectivity into World Society: The Case of Extending Personhood to Plants." *Modern Subjectivities in World Society: Global Structures and Local Practices,* edited by Dietrich Jung and Stephan Stetter. Macmillan, 2019, pp. 211–227.

Rieger, Stefan. "What's Talking?: On the Nostalgic Epistemology of Plant Communication." *The Green Thread: Dialogues with the Vegetal World,* edited by Patrícia Vieira, Monica Gagliano and John Ryan. Lexington Books, 2016, pp. 59–79.

Ryan, John Charles. "Passive Flora? Reconsidering Nature's Agency Through Human-Plant Studies (HPS)." *Societies,* vol. 1, 2012, pp. 101–121. doi:10.3390/soc2030101.

———. "Planting the Eco-Humanities? Climate Change, Poetic Narratives, and Botanical Lives." *Rupkatha Journal on Interdisciplinary Studies in Humanities,* vol. 8, no. 3, 2016, pp. 61–70. rupkatha.com, http://dx.doi.org/10.21659/rupkatha.v8n3.08.

Sandilands, Catriona. "Plants." *The Cambridge Companion to Environmental Humanities,* edited by Jeffrey Jerome Cohen and Stephanie Foote. Cambridge University Press, 2021, pp. 156–169.

Vernon, Zackary. "Faulkner's Charismatic Megaflora: Critical Plant Studies and the US South." *Journal of Modern Literature*, vol. 45, no. 3, 2022, pp. 89–105. *Project Muse*, http://dx.doi.org/10.2979/jmodelite.45.3.07.

Wandersee, James H., and Elisabeth E. Schussler. "Preventing Plant Blindness." *The American Biology Teacher,* vol. 61, no. 2, 1999, pp. 82–86. https://www.jstor.org/stable/4450624.

Chapter 21

Gandhian Ecosophy

Reflections on Ecology and Self

Narayan Jena

Akeel Bilgrami, in "Gandhi, the Philosopher" (2003), has called Gandhi an "avowed humanist" (4162). Bilgrami's expression is a reiteration of the phrase uttered by the scholars like Aldous Huxley, C. F. Andrews, Gilbert Murray, Anand Coomarswamy, Arun Gandhi, Sarvepalli Radhakrishnan and many more. This expression has topical reference to Gandhi's ideals, which are based on his deep engagement in humanitarian practices and religious beliefs. Gandhi's ideals were the instruments to chisel his political thoughts, which makes him an astute nationalist leader. John Haynes Holmes has pointed out his humanistic attitude by highlighting his unflinching toil to revive the sense of "native cultural identity," "personal dignity or self-respect" and *Swaraj* (home rule or political independence) (75). It is evident that such humanistic expression goes with strong spiritual and political intents. Gandhi, as Radhakrishnan has observed, uses spiritualism and politics as the modalities to explore the "truth" and the "self," and to fight against the forces that minimise the possibilities of unfolding the "truth" and realisation of "self" (7–8). This chapter is a humble attempt to foreground and examine the ecological implications of Gandhi's humanistic philosophy. His humanistic views, which exemplify his altruistic and egalitarian nature, are not constricted to the human world; rather, they include the whole universe and celebrate the innate goodness of all beings. Gandhi's ecosophy as a set of ecological concepts, which emanates from his austere spiritual practices and deep respect for all *life forms* of nature, is the testimony to his profound understanding of the fundamental relationship of human life with the nonhuman world.

Gandhi as a humanist disengages himself from the "self-enclosed" subjects. In his humanistic view, boundaries of "self" expands to include the non-human "other." His incessant endeavour to minimize his sense of "self-importance" and to dissociate himself from material attachments suffice to prove him a spiritualist who truly cares for all beings. Although Gandhi's humanistic and religious practices are truly ecological, he has not postulated any distinct environmental theories or ecological ideas to prove himself as a systematic ecological thinker. However, the fundamental practices in his life unfold a set of practical philosophies which unveil the spiritual dimensions of "self" and its relationship with the world. So, to understand Gandhi's philosophy, it is necessary to have a clear understanding of the layers of meaning embedded in his life and living practices. This chapter attempts to critically interpret certain principles and practices of Gandhi to understand their deeper connection with environmentalism and ecophilosophy. It will show how Gandhi's philosophical ideas, based on Hindu *advaita tradition* and Jain practices, share certain areas of commonality with the philosophy of Arne Naess, yet they are distinct.

It is true that ecophilosophy is quite a recent postulation, pioneered by the Norwegian philosopher Arne Naess. Gandhi came much before the postulation of the *deep ecological strands* of environmental philosophy. As an antecedent to ecophilosophy, his name is less known in the field of deep ecology. His name comes at the end of the list that comprises the leading thinkers like Allan Watts, Deisetz Suzuki, Gary Snyder, Carolyn Merchant and others. However, Naess has admitted that he was "inevitably influenced by Gandhian metaphysics" (qtd. in Drengson and Devall 25), especially by the ideas of "self-realization," Truth, *ahimsa* (non-violence), simple living with minimum possession, etc. Such ideas possibly influenced Naess's "biospherical egalitarianism" (Naess, "Self-Realisation" 236) and promoted the ideas of holistic living in harmony with nature through the identification of the "self" as the part of the world. Gandhi's idea of "non-possession of material abundance" as the method to let others live with all rights and dignity imbued the ecosophical idea of the "maximum unfolding" of being through the self-realization of all (Weber 351). Such a sense of deep-seated respect for all forms of life constitutes the locus of ecophilosophy.

Naess's ecosophy is, therefore, defined as "a philosophy of ecological harmony or equilibrium" (Drengson and Inoue 8). This wisdom is a set of normative principles, makes us understand the "living bond between ourselves and the rest of the animate world" (Patel and Sykes 50). In the Introduction of Naess's *The Ecology of Wisdom* (2008), Drengson mentioned that this philosophical knowledge "examines our basic values and lifestyle and reflects on our fundamental relationships with nature and who we are" (27). In an ecological understanding, every individual being (human and non-human) is a part of the complex biospheric system. Nature as a living system provides necessary conditions for the existence and flourishing of beings as well as determines "who we are in the larger scheme of things" of nature (Weber 352). In other words, ecophilosophy looks into the complexity of relationships among all beings who exist in an interdependent, interconnected organic community. At a deeper level, it looks into the process of "self-realization" of being with relation to the greater "Self" of nature. If self-realization is a process of attaining maximum "unfolding" in an ecophilosophical sense, Gandhi's self-realization is concerned with the sacredness of all beings and their connection with God. Gandhian "self," which is altruistic in nature, recognizes every single organism as significant as the other in the process of "self-realization" of all beings.

Gandhi perceives that *all life is of one piece*. He further muses that "to see the universal and all-pervading spirit of Truth face to face, one must be able to love the meanest of creations as oneself" (qtd. in Radhakrishnan 2). It shows Gandhi's reverence for all beings because, to him, all life is divine. In Gandhian sense, the beings that live and the world that gives them necessary condition to live and flourish are divine and work in a continuum. In such a reverential continuum, all things are intrinsically connected to *Brahman*, or the universal Self. In an ecological sense, it is what Naess calls "Life together" or "the existence of interdependencies" in which all beings get enriched and evolved (Naess, *Ecology, Community and Lifestyle* 203). Although evolution is inevitable in nature and culture, it does not occur in Darwinian way—through constant conflict. Gandhi, like Naess, argues that evolution happens through interface and interdependence. This idea is ecologically valid because, in a system where everything depends on everything else, nothing can be more or less qualitatively significant than anything else. Through interdependency, *the finest of the being* can be given maximum unfolding.

Gandhi's egalitarian views of life forms as "having equal value" (Gier 30) minimises the differentiation among all beings. The *spiritual energy* that permeates every individual being elevates the human and non-human, the living and non-living, to the same level.

In Gandhi's philosophy, the beings, which are intrinsically connected to *Brahman* or the God or what Naess calls "the Universal Self," are not independent. In the continuum, they are connected through the same *spiritual energy*. The material principles of the world are derived from this energy and nature's material forms are the embodiment of such energy. Gandhi says, "There is an indefinable, mysterious power that pervades everything. I feel it though I do not see it" (qtd. in Radhakrishnan 3). He further upholds that his existence is possible without basic needs but not "without Him." This "life force" or *Brahman,* in Gandhi's belief, is *non-dual* or *advaita* which does not have any form other than the "absolute" which in Gandhian sense is the highest "Truth." It is evident that Gandhi's *advaita* God is similar to, yet different from Spinoza's *monistic God*. Gandhi's God is not dissociated from the beings; rather God as the "all-encompassing cause" expresses itself in beings in different forms present in the "reverential continuum." Gandhi's God or *Brahman* is non-dual, but not isolated and transcendental. In other words, the boundaries between the divine and the ordinary, the ideal and the real, do not separate them as distinct categories. So, when Gandhi's non-dual God is immanent and present in all beings as the cause, Spinoza's monistic God is immanent but transcendent. In Spinoza's "substance monism," God is the substance with different attributes and appellations. If substance is the essence of being, the being is dependent on the substance and nothing else. This being is like *jiva* in Jain Philosophy which is "self-sufficient" and distinct from other beings (Gier 32). But Gandhi's "self," which is nothing but the "cause" in itself, exists in consonance with other beings in an environment from which the being derives its context and meaning. Again, Gandhi's idea of *Daridranarayan* (may be translated as the God of the poor) (*India of My Dream* 53) suggests that his God is not majestic or sovereign, who is controlling all beings from the transcendental point. The God, unlike Platonic God, is one among the multitude. He understands the existential necessities of the beings. Anything that fulfils such necessities is divine. For Gandhi, "God can only appear as bread and butter" (*India of My Dream* 54). So, his God appears to be so mundane that the multitudes of beings do not recognise his presence. He says, "I worship the God that is truth or Truth which is God through the service of these millions" (*India of My Dream* 54).

It is evident in his writing that Gandhi has often resorted to *advaita* philosophy. He has admitted that he is a devout follower of "*advaita* doctrines" (*CWMG* 165). He says, "I believe in *Advaita*, I believe in the essential unity of man, and for that matter of all that lives. Therefore, I believe that if one man gains spirituality, the whole world gains with him, and if one man falls, the whole world falls to that extent" (390). According to Margaret Chatterjee, Gandhi follows *advaita* to the extent of his belief in the "oneness of all that lives" (105). Similarly, Nicholas D. Gier says, "Gandhi declares allegiance to *Advaita Vedanta*, but he only interprets it to mean the unity of God and humans" (Gier 38). His idea of *varnashramadharma* repudiates any kind of superiority and inferiority among human beings. As a humanist, his understanding of *advaita* is oriented towards human, and his God is not the *advaita* God in the classical sense. And his understanding of *advaita* does not explicitly go in compliance with the fundamental principles of *Advaita Vedanta*. However, Gandhi's "self" and "God" can very well be compared with Naess's "self" and "Self" (with "S").

In Naess's philosophy, "Self" permeates everywhere, and everything is the extension of the "Self." Naess's "Self" is both the cause and the substance;

it is the life force that expresses itself in both the "system" and the "being." Naess, in his writing, has frequently used two self(s)—the first one with "S" to suggest broad and universal self or "ecological self" and the second with "s" to suggest the individual or ego self ("Deep Ecological Movement" 82). Ego is a construct, the supercilious impression created by the individual about himself or herself through their consciousness, memories, experiences, etc. To Bill Devall, ego-self is "a collection of memories, fantasies, information and images about who we are. It is what we think we are, not what we experience as our self" ("The Ecological Self" 103). Naess, like Gandhi suggests that an individual self exists with relation to its community, bioregion, environment, and other living and non-living beings. In this sense, the "other" (as different from and inferior to "self") appears to be redundant because the "other" is not the other in a generic sense as it is an essential constituent of a system that operates in correlation. If every individual in a system has equal importance, it is impossible to determine which the self (in the centre) is and which the other (in the margin) is. If ego is an imaginary boundary that confines the self with "maturity," the self makes move from "ego self" to "Ecological self."

Gandhi believes that self-realisation is a process of attaining the highest level of spirituality through the realisation of the *Brahman* in all beings. Realisation of self is a process through which the being transcends the "confinement of petty selves" (Coomarswamy 49). The being that is obsessed with self-love fails to acknowledge the value of other beings. In Gandhian sense, to realise the "greater Self" or Brahman, it is imperative to realise "the spiritual essence of all that is" (48). From Arne Naess's perspective, the self that expands itself to recognize the "spiritual essence" in all is the "Ecological self." In Devall's understanding it is the "widened and deepened self" with an elevated consciousness that "there is an identification of all life" ("The Ecological Self" 108). So, "Ecological self" is not a forced attribute or a point to be attained, but a process, in which the "self" explores and realises its relation with the living world. Such realisation is the part of an evolutionary journey in which develops a sense that the whole living world emanates from the "Self" and the individual self exists in integration with others. Devall further posits that cultivation of Ecological self means learning "how to joyfully blend with the watershed in which we live" (109). Such sense of blending with the larger scheme of things, in both Gandhi and Naess's philosophy, refers to infinite task of "Self-realisation." For that matter, identification with nature is the modality for Self-realisation, the process by which "ecological self" is experienced. It implies that development of *ecological self* is the consequence of Self-realisation. And the consequence is a move along with the process because we know that "ecological self" is not a fixed destination as is "self-realisation." In a Gandhian sense if a "universal force or energy" function everywhere as the prime cause of all material expressions, this energy (as *Atman, Brahman, Supreme Being, God, Life Force, etc.*) causes the "capacity of identification." If extension of identification to the ecological self is a move towards self-realisation and extension of self to identify everything as the part of it, the sense of self-realisation and ecological self share a common ground, that is the opening of the boundary of the "self." It is quite evident that in both Naess and Gandhi's philosophy, there is no indication of the abnegation and annihilation of individual "narrow self"; rather there is the assertion for the elevation of the "self" from the confinements of "egocentric interest" to the vast openness of the world. That is Moksha for Gandhi or *jeevan mukta*.

Gandhi's *jeevan mukta* (liberated life) exemplifies the emancipation of the self from the "egocentric interests" through the dissociation from material clinging. The practice of non-acquisition of material wealth beyond necessity is similar to Jain practice of *aparigraha. Aparigraha* involves the practice of limiting one's needs to the minimum and complete renunciation of one's greed. Greed is an illegitimate practice of stealing other's rights over nature's resources. As Gandhi rightly said, the resources of nature are enough to cater to the needs of all beings and they are highly insufficient to satisfy the greed of humanity. So, in the democratic system of Earth, the non-possession of "the maximum" is necessary to ensure the rights of all beings over the resources in order to live and flourish. Greed, according to Gandhi, is "the vehicle for exploitation of others" (*India of My Dream* 68). Gandhi has suggested "the contentment for all" which is possible through the elevation of the living qualities of all. Living quality enhances through "living at peace with the world" and "learning the art of living nobly" (*India of My Dream* 31) and not through maximum material accumulation. Noble life, according to Gandhi, is the life of universal benefit, which is possible through the "maximization of giving away." Gandhi posits that greed leads to intolerance and violence; it minimises the possibility of others to exist and unfold. Intolerance does not let others remain "other," rather the "other" is conditioned to become the part of one's "ways of life." It leads to "an inclusionary homogenising attitude" persuaded with "violence towards the other" (Bilgrami 4163). Violence brings about a complete decimation of "other(s)." As a result, it is a threat to the world order. So, Gandhi in his philosophy has resorted to *ahimsa,* which should be translated as "love" (Andrews X).

Ahimsa (non-violence), for Gandhi, is not "a mere philosophical principle," it is "the rule" of his life (Andrews 95). It is widely known that Gandhi combined 'non-violence' with his political intelligence to devise the most powerful instrument of *satyagraha* (war for the Truth) to fight against British oppression. But the true intention of Gandhi's non-violence was to spread love and to bring about complete freedom of all living beings from all kinds of domination and violence. His *ahimsa* is based on the philosophy of *advaita*, the non-dual God and the idea of "universal kinship of all souls" (Gier 30). *Ahimsa* develops the fundamental understanding that all living beings are equal, and they are like "one's own self." *Himsa* (violence) and oppression in any form violate the ethical norms of humanity. But absolute negation of violence is impossible because "all life in the flesh exists by some violence" (Andrews 91). This idea has been endorsed by Naess that for sustenance a being always depends on other beings. Gandhian Philosophy makes it clear that resorting to violence with intentions to eliminate "others" in order to fulfil self-greed is real violence; thus, violence that decimates the existence of others is ethically and environmentally wrong. As ambition and greed are the causes of violence, and "oppression of the weak," (Andrews 90) Gandhi has insisted on setting "limit to our worldly ambition" and maximizing "godly pursuits" (*Hind Swaraj* 34). Godly pursuits, in ecological terms, refers to the belief in "the sacredness" and "interconnection" of all life forms in the "reverential continuum." And the highest sense of non-violence develops a sense of identification of "oneself to be a genuine part of all life" (*Ecology, Community and Lifestyle* 174). In the process of identification with others, the atomistic self-ego loses its strength, and every living being is "understood as a goal in itself" and as a living possibility. Thus, infliction of violence on others is self-damaging. Non-violence generates a truly egalitarian attitude in which all life matters.

Gandhi's reverence for all forms of life is exemplified in his cow worship and reverence for "mother earth" which are the metaphors for expressing unconditional love for the non-human world. Gandhi posits, "The cow to me is the entire sub-human world" (*India of My Dream* 130). Although cow protection and worship are associated with Hindu religious practices, it is "intimately connected with the culture of the soil" (Andrews 5). In an agrarian society like India, cow is not a mere animal. The domestication of cows adds economic value to households. According to Gandhi, cows not only give milk but also "makes agriculture possible." This indicates the dependence of humans on the non-human. Gandhi says, "I myself respect cow, I look upon her with affectionate reverence" (*Hind Swaraj* 41). He further muses with gratitude that cow is "the mother to millions of Indian mankind. Protection of the cow means protection of the whole dumb creation of God" (*India of My Dream* 130). In Gandhian sense, cow protection is not an irrational act, rather it has both religious and economic significance. The religious meaning of cow is derived from the material purpose of cow protection. Again, Gandhi's reverence for the world continues to be expressed in his "mother earth" metaphor. Regarding Earth as the mother and expressing deep sense of reverence is the part of Hindu agrarian practices in India. Such reverence stems from our gratitude for Earth's procreating and life sustaining potential. For its procreative nature, in Gaia hypothesis, James Lovelock considers Earth as a living creature, and to Thomas Berry, it is the "spiritual energy" that continues to "act in all that acts upon the earth" (*The Sacred Universe* 69). Similarly, Ned Hettinger reasserts Earth's sacredness for its magnificence of physical structure that supports "miraculous earthen community of life process" ("Ecospirituality" 83). Gandhi understands our lives as a part of that living community.

Gandhi's deep concern for the world is expressed in his practices of "simplicity" in life. His practice of deriving maximum contentment from limited material acquisition ensures maximum material provisions for the existence of other beings. Gandhi's idea of mutual enrichment is similar to Naess's "deep ecological lifestyle." To Gandhi, enrichment of one's life is impossible without enriching others. Similarly, Naess says that the quality of life of a human being is determined by the quality of life of other species in a collective way. So, both Gandhi and Naess suggest a frugal and simple life, which can be the modality to enrich the self by ensuring a good quality of life for others.

Gandhi's simplicity of lifestyle is an indelible message for sustainable development. Gandhi advocated for transforming society through the sensible use of resources, which entails the process of development without exploiting the resources of future generations and the underprivileged. Gandhi's model of development is conditioned by self-defined local goals. So, Gandhi's models are not unidirectional, but rather aim at addressing local problems like poverty, socio-economic disparity, and environmental insecurity. To Gandhi, economic equality is the key to harmonious relationships in society. He states that economic equality means abolishing the "eternal conflict between capital and labour." To bring social and ecological exploitation to an end, the socio-economic conditions of people need to be elevated. This is the reason why Gandhi's focus is on the development of rural India where he witnessed maximum poverty. Rural development is not to change the pristine countryside into an industrial hub or to develop massive industrial structures at the cost of rural resources; rather, it is to develop an economy in which the lives of people and non-human nature matter. So, Gandhi has advocated for a developmental model that can promise economic welfare for all, and a good quality of life, instead of massive material growth.

In this context, Gandhi's rejection of the Western industrial model of development appears to be quite judicious. The Western models seem to be inappropriate in the Indian environment because they focus on production rather than people's economy. For that matter, the massive industrial structures are not suitable where there happens to be a massive economic disparity. And the idea of decentralised economy is possible when production is "controlled by the mass" (*India of My Dream* 72) rather than by the capitalists. Gandhi's ideas of cottage industries, cooperative farming, and small projects can be the alternatives to the massive industries in India. And such industries are ecologically viable and can secure healthy and dignified life for the masses. Gandhi's idea of minimization of the use of machines aims at giving maximum employment to the underprivileged, through which economic welfare of the last man in the social hierarchy is possible.

Gandhi's ecophilosophy is not limited to any time and location. As an eco-concept it has global implications. It imbues both episteme and practices concerning global ecology. His ideas of *ahimsa* (non-violence) and *aparigraha* (non-possession) are the ethical guidelines for the nations which believe in the philosophy of development through massive technological growth. His concept of development as a practice of voluntary reduction of wants can be viewed as the modality to facilitate sustainable development by controlling human greed. Such practice also includes the concern for other beings and the development of all. Gandhi's repudiation of the developmental model that inflicts irreversible ecological damage and induces socio-economic disparity is a way to promote global ecological justice and the economy of permanence. Such ecological idealism, expressed through his cardinal principles practiced in his *ashram life* at Shabarmati, can reorient the global environmental discourses towards the philosophy that advocates for the ethical growth of all beings. However, it is undeniably true that we are in a time when the world is troubled by rising environmental degradation, and technology has failed to give any permanent solution to such problems. When the existing technological and managerial approaches are inappropriate to mitigate the problems, we need to turn towards a philosophy which can bring about changes in human behaviour and consciousness. In this context, Gandhi's humanistic philosophy that focuses on the quality of life in *harmony* can potentially reorient the human mind and develop "ethical sense" towards nature. Gandhi's life and his philosophical message need to be appreciated, analysed and followed to find possible solutions to the existential problems of our time.

REFERENCES

Andrews, Charles F. *Mahatma Gandhi: His Life and Ideas.* Jaico Publishing House, 2015.

Berry, Thomas. *The Sacred Universe: Earth, Spirituality and Religion in the 21st Century.* New York: Colombia University Press, 2009.

Bilgrami, Akeel. "Gandhi, the Philosopher." *Economic and Political Weekly,* vol. 38, no. 39, 2003, pp. 4159–4165.

Chatterjee, Margaret. *Gandhi's Religious Thought.* University of Notre Dame Press, 1983.

Coomarswamy, Anand K. "Mahatma." *Mahatma Gandhi: Essays and Reflections*, edited by Sarvapalli Radhakrishnan, Jaico Publishing House, 2015, pp. 46–48.

Devall, Bill. "The Ecological Self." *The Deep Ecology Movement: An Introductory Anthology,* edited by Alan Drengson and Yuichi Inove, North Atlantic Books,1995, pp. 101–123.

Devall, Bill and George Sessions. *Deep Ecology: Living as if Nature Mattered.* Gibbs Smith, 1985.

Drengson, Alan and Bill Devall. Introduction. *Ecology of Wisdom: Writings by Arne Naess,* edited by Drengson & Devall. Counter Point, 2008, pp. 1–49.

Drengson, Alan and Yuichi Inoue, editors. *The Deep Ecology Movement: An Introductory Anthology.* North Atlantic Books, 1995.

Gandhi, Mohandas K. *The Collected Works of Mahatma Gandhi (CWMG).* Publication Division, Government of India, 1969.

———. *Hind Swaraj.* 7th ed. Rajpal and Sons, 2016.

———. *India of my Dream.* 9th ed. Navjivan Trust, 2016.

Gier, Nicolas F. *The Virtue of Non-Violence: From Gautam to Gandhi.* State University of New York Press, 2004.

Hettinger, Ned. "Eco Spirituality: First Thoughts." *Dialogue and Alliance: A Journal of the Instructional Religion Foundation*, vol. 9, no. 2, 1995, pp. 81–98.

Holmes, John Haynes. "The Nature of Gandhi's Greatness." *Mahatma Gandhi: Essays and Reflections*, edited by Sarvapalli Radhakrishnan. Jaico Publishing House, 2015, pp. 74–76.

Lovelock, James. *The Revenge of Gaia.* Penguin Books, 2006.

Naess, Arne. *Ecology, Community and Life Style: Outline of an Ecosophy.* Translated by David Rothenberg, Cambridge University Press, 1989.

———. "Deep Ecological Movement: Some Philosophical Aspects." 1986. *Deep Ecology for the 21st Century: Readings on the Philosophy and Practice of the New Environmentalism*, edited by George Sessions. Shambhala Publications, 1995, pp. 64–84.

———. Self-Realisation: An Ecological Approach to Being in the World." *Ecology for the 21st Century: Readings on the Philosophy and Practice of the Environmentalism*, edited by George Sessions. Shambhala Publications, 1995, pp. 225–240.

Patel, Jehangir P. and Marjorie Sykes. *Gandhi: His Gift of the Fight.* Friends Rural Centre, 1987.

Radhakrishnan, Sarvapalli. "Gandhi's Religion and Politics." *Mahatma Gandhi: Essays and Reflections*, edited by Sarvapalli Radhakrishnan. Jaico Publishing House, 2015, pp. 1–26.

Weber, Thomas. "Gandhi, Deep Ecology, Peace Research and Buddhist Economy." *Journal of Peace Research,* vol. 36, no. 3, 1999, pp. 349–361.

Wilde, Oscar. "Panthea." Kalliope, 2023, https://kalliope.org/en/text/wilde2001070750. Accessed March 20, 2023.

Index

About the Editors and Contributors

EDITORS

Cenk Tan received his BA in American culture and literature from Hacettepe University (Ankara, Turkey) in 2002. He earned his PhD from the Department of English Language and Literature. Tan is a trilingual speaker of English, French, and Dutch and specializes on topics such as science fiction, ecocriticism, cinema/film criticism, and continental philosophy. He works as a lecturer doctor at Pamukkale University (Denizli, Turkey) and is the author of various scholarly articles. He is also the editor of many published and upcoming books. Tan is currently co-editing the upcoming volume *Class Conflict in 21st Century Science Fiction Film.*

İsmail Serdar Altaç earned his PhD from the Department of English Language and Literature at Ankara University with his dissertation titled 'The Representation of Urban Space in James Graham Ballard's Novels'. He currently works as an assistant professor in the Department of English Language and Literature at Nevşehir Hacı Bektaş Veli University, Turkey. His research interests include contemporary fiction, dystopia, utopia, and ecocriticism.

CONTRIBUTORS

Dr. **Panchali Bhattacharya** is assistant professor (English) in the Department of Humanities and Social Sciences, National Institute of Technology Silchar, Assam, India. She has completed her PhD from the School of Humanities, Social Sciences, and Management, IIT Bhubaneswar. Her research interests include ecocriticism, myth and folklore studies, and indigenous literature with a special emphasis on the Northeastern region of India.

Brian Deyo is associate professor of English at Grand Valley State University in Allendale, Michigan, where he teaches courses in world literature and the environmental humanities. His scholarship in postcolonial ecocriticism has been published in journals such as *ariel: A Review of International English Literature* and *ISLE:* Interdisciplinary Studies in Literature and the Environment.

Dr. **Erden El** earned his BA in English language and literature from Ankara University. After completing his MA at Atılım University, Department of English Language and Literature, El received his doctorate from Hacettepe University, Department of American Culture and Literature. El works as an assistant professor at Ankara Social Sciences University. He has published articles and chapters on ecocriticism. El is generally interested in the environmental justice theme of the theory of ecocriticism. He also works on video games and ecocriticism.

Lenka Filipova completed her doctorate in English studies at Freie Universität Berlin. Recent publications include *Ecocriticism and the Sense of Place* (2022), and contributions to *The Routledge Handbook of Ecofeminism and Literature* (2022) and *The Routledge Handbook of Trans Literature* (2024). She is currently at work on her second book project, which explores 'material fictions' in literary and cultural narratives in the global eighteenth century.

Mahinur Gözde Kasurka completed her PhD in English literature department at Middle East Technical University with her dissertation titled 'A Posthumanist Study of the Dystopian Novel: Margaret Atwood's Oryx and Crake, Jeanette Winterson's The Stone Gods, David Mitchell's Cloud Atlas'. Currently, she works as a full time instructor doctor at National Defence University, School of Foreign Languages. Also, she works as a part time scholar at İstanbul Bilgi University. Her research interests are British novel, literary theory, women and writing, psychology and literature, dystopian fiction, ecocriticism and posthumanism. She has presented several papers at various international conferences. She has attended Rosi Braidotti's summer school on 'Posthuman Knowledge(s)'.

Narayan Jena is assistant professor in the Faculty of Arts, Communication and Indic Studies at Sri Sri University, India. His areas of interest include ecophilosophy, postmodern ecology, postcolonial ecocriticism, etc. He is a trained ecocritic and an admirer of Arne Naess's ecosophy T.

Philippe Lynes is Addison Wheeler fellow in the Department of English Studies at Durham University and held the 2017–2018 Fulbright Canada visiting research chair in environmental humanities at the University of California, Irvine. His research situates itself at the intersections of the environmental humanities, continental philosophy, and ecocriticism.

Dr. **Amarjeet Nayak** received his PhD at the IIT Kanpur in 2009, after which he has served as a faculty member at Thapar University, Patiala, and the Indian Institute of Technology Indore. Currently, he is a reader in the School of Humanities and Social Sciences at the National Institute of Science Education and Research Bhubaneswar, an OCC of Homi Bhabha National Institute, India. He has published his papers in many reputed international and national refereed journals and anthologies. His primary areas of interest are postcolonial studies, speculative fiction studies, film studies, and Indian English literature. He is also a bilingual writer who writes short stories in Odia and English.

Ufuk Özdağ is professor of American culture and literature and founder-director of the Land Ethic Research and Application Center at Hacettepe University, Ankara, Turkey. Özdağ specializes in American nature writing, ecocriticism, and comparative studies of the environmental literatures of the United States and Turkey. Özdağ is the author of *Literature and the Land Ethic: Leopoldian Thought in American Nature Writing* (2005) and *Introduction to Environmental Criticism: Nature, Culture, Literature* (2014), both in Turkish. She is the co-editor of *The Future of Ecocriticism: New Horizons* (2011), *Environmental Crisis and Human Costs* (2015), *Anadolu Turnaları: Biyoloji, Kültür, Koruma* (*Anatolian Cranes: Biology, Culture, Conservation*) (2019), and *Nature v/s Culture* (2022). Özdağ is the translator of Aldo Leopold's conservation classic *A Sand County Almanac* (2013), and the co-translator of Edward Abbey's seminal work *Desert Solitaire* (2019) into Turkish. Özdağ initiated and has been offering 'literature and environment' and 'environmental studies' courses at Hacettepe University for fifteen years. As a 'land ethic leader' (Aldo Leopold Foundation's LEL graduate 2010), Özdağ is pioneering efforts in land restoration at Hacettepe's Beytepe Campus.

Dr. **Pritam Panda** is assistant professor in the Department of English, Jogananda Deva Satradhikar Goswami College, Bokakhat, Assam, India. He received his PhD from the University of Lucknow, India. Dr. Panda completed his masters degree in English literature from Utkal University,

Bhubaneswar, Odisha. His research interests include science fiction, speculative literature, sports literature and film studies.

Dr. **Abhra Paul** is postdoctoral fellow at the National Institute of Science Education and Research (NISER), Bhubaneswar, an OCC of Homi Bhabha National Institute, India. She received her PhD from the Indian Institute of Technology, Jodhpur, in 2021. Her research areas include English literature, American literature, South Asian literature, culture studies, blue humanities, the Anthropocene, and environmental humanities.

Denis Petrina (he/him) received his PhD in philosophy from the Lithuanian Culture Research Institute (Vilnius, Lithuania). His doctoral dissertation explores philosophical interpretations as well as (bio)political implications of the notion of affect. His publications mostly focus on affect theory in the contexts of both classical and contemporary philosophy, (re)interpretations of biopolitics, and, more recently, digital governmentality in cybercapitalism. His research interests include affect theory, theory of subjectivity, biopolitics, media studies, sexuality and queer studies, speculative epistemologies. He is a part of a research cluster 'Immunity and Contagion: Transformations of Biopolitics in the Time of a Pandemic' (2021–2024, funded by the Research Council of Lithuania).

Ziba Rashidian teaches courses on animal studies, the environmental humanities and comparative literature with a focus on trauma and cultural memory. She edited, along with Jeanne Dubino and Andrew Smyth, *Representing the Animal in Modern Culture* (2013). Her current research interests focus on the intersection of human and nonhuman animal lives at moments of environmental catastrophe.

Ana Simić (Sentov) is associate professor at the Department of English, Faculty of Law and Business Studies Dr Lazar Vrkatić. She has taught English literature, English-Serbian translation, English for specific purposes, and intercultural communication. Her research interests include postmodernist and postcolonial literature in English, ecofeminist and ecological criticism, and translating culture-specific content.

Dr. **Milena Škobo** is assistant professor at the Department of English Language and Literature at Sinergija University, Bosnia and Herzegovina, where she teaches diverse courses in English and American literature. Her research focuses on fictional semantics, eco-narratives, environmental literature, and methodology of teaching. She is passionately committed to raising awareness about the crucial role of literature in fostering critical thinking, independent reflection about the world and social change.

Clara Soudan completed her PhD in political philosophy at the University of Edinburgh. Her doctoral dissertation studies the ecological crisis from the prism of Gnosticism and the motive of a dualistic alienation from the world. Delving into questions of intimacy and liminality in our relationship with more-than-humans, her research explores more broadly the rediscovery of our ontological entanglement with more-than-human life and its impact upon modern systems of thought.

Gina Stamm is assistant professor of French at The University of Alabama. Their research focuses on the relationship between the body and its environment in French and francophone avant-gardes from surrealism to contemporary science fiction, primarily through the lenses of ecocriticism and gender theory.

Dr. **Inna Sukhenko** is research fellow of Helsinki Environmental Humanities Hub, Department of Cultures, University of Helsinki. Her current project is focused on researching the literary dimensions of nuclear energy within energy literary narrative studies. After defending her PhD in Literary Studies (Dnipro, Ukraine), she has been a research fellow of Erasmus Mundus mobility programmes (Bologna, 2008; Turku, 2011–2012), Cambridge Colleges Hospitality Scheme (2013), Open Society Foundation/Artes Liberales Foundation (Warsaw, 2016–2017), JYU Visiting Fellowship Program (Jyväskylä, 2021), and PIASt Fellowship Program (Warsaw, 2021). Her general research interests lie within environmental humanities, energy humanities, ecocriticism, nuclear criticism, literary energy narrative studies, world energy literature, nuclear fiction, and Chernobyl fiction.

Rebekah A. Taylor-Wiseman, PhD, currently writes proposals and applies interdisciplinary frameworks to develop solutions for public and environmental health problems at Banyan Communications, a cause-driven creative consultancy in Atlanta, Georgia. She is a former professor of English and writing center director who is holding out hope for collaborative, inclusive actions to improve health equity and environmental justice in communities and globally. She also has the privilege of being a mom to three wild, magical children who are teaching her how to practice radical love and kindness.

Aina Vidal-Pérez (PhD comparative literature) is postdoctoral fellow at the Universitat de les Illes Balears. She is a member of the R&D project 'The Novel as Global Form. Poetic Challenges and Cross-border Literary

Circulation' and collaborates in the ERC Consolidator Grant project 'Ocean Crime Narratives: A Polyhedral Assessment of Hegemonic Discourse on Environmental Crime and Harm at Sea (1982–present)'. Her research interests focus on systemic approaches to world literature, the global novel, environmental criticism and Mediterranean studies.

Kübra Vural Özbey is assistant professor of English in the Department of Western Languages and Literatures at Muğla Sıtkı Koçman University (Muğla, Turkey). She completed her doctorate in 2021 at Hacettepe University (Ankara, Turkey) after conducting PhD research at the University of Illinois at Urbana-Champaign as a Fulbright grantee from January to October 2021. Her recent publications include 'A Rite of Passage in *Shakespeare's Love's Labour's Lost*: 5.2.786-806' (*The Explicator*, 2023) and 'Richard III: Why I Did It (Review)' (*Shakespeare Bulletin*, 2023). Vural Özbey's research interests are Shakespeare, Irish drama, literary theories and criticism.

www.ingramcontent.com/pod-product-compliance
Ingram Content Group UK Ltd.
Pitfield, Milton Keynes, MK11 3LW, UK
UKHW041835031026
15587UKWH00002B/776

9 781666 923483